BOOLEAN MATRIX THEORY AND APPLICATIONS

PURE AND APPLIED MATHEMATICS

A Program of Monographs, Textbooks, and Lecture Notes

MONOGRAPHS AND TEXTBOOKS IN PURE AND APPLIED MATHEMATICS

1. *K. Yano,* Integral Formulas in Riemannian Geometry (1970) *(out of print)*
2. *S. Kobayashi,* Hyperbolic Manifolds and Holomorphic Mappings (1970) *(out of print)*
3. *V. S. Vladimirov,* Equations of Mathematical Physics (A. Jeffrey, editor; A. Littlewood, translator) (1970)
4. *B. N. Pshenichnyi,* Necessary Conditions for an Extremum (L. Neustadt, translation editor; K. Makowski, translator) (1971)
5. *L. Narici, E. Beckenstein, and G. Bachman,* Functional Analysis and Valuation Theory (1971)
6. *D. S. Passman,* Infinite Group Rings (1971)
7. *L. Dornhoff,* Group Representation Theory (in two parts). Part A: Ordinary Representation Theory. Part B: Modular Representation Theory (1971, 1972)
8. *W. Boothby and G. L. Weiss (eds.),* Symmetric Spaces: Short Courses Presented at Washington University (1972)
9. *Y. Matsushima,* Differentiable Manifolds (E. T. Kobayashi, translator) (1972)
10. *L. E. Ward, Jr.,* Topology: An Outline for a First Course (1972) *(out of print)*
11. *A. Babakhanian,* Cohomological Methods in Group Theory (1972)
12. *R. Gilmer,* Multiplicative Ideal Theory (1972)
13. *J. Yeh,* Stochastic Processes and the Wiener Integral (1973) *(out of print)*
14. *J. Barros-Neto,* Introduction to the Theory of Distributions (1973) *(out of print)*
15. *R. Larsen,* Functional Analysis: An Introduction (1973)
16. *K. Yano and S. Ishihara,* Tangent and Cotangent Bundles: Differential Geometry (1973)
17. *C. Procesi,* Rings with Polynomial Identities (1973)
18. *R. Hermann,* Geometry, Physics, and Systems (1973)
19. *N. R. Wallach,* Harmonic Analysis on Homogeneous Spaces (1973)
20. *J. Dieudonné,* Introduction to the Theory of Formal Groups (1973)
21. *I. Vaisman,* Cohomology and Differential Forms (1973)
22. *B.-Y. Chen,* Geometry of Submanifolds (1973)
23. *M. Marcus,* Finite Dimensional Multilinear Algebra (in two parts) (1973, 1975)
24. *R. Larsen,* Banach Algebras: An Introduction (1973)
25. *R. O. Kujala and A. L. Vitter (eds),* Value Distribution Theory: Part A; Part B. Deficit and Bezout Estimates by Wilhelm Stoll (1973)
26. *K. B. Stolarsky,* Algebraic Numbers and Diophantine Approximation (1974)
27. *A. R. Magid,* The Separable Galois Theory of Commutative Rings (1974)
28. *B. R. McDonald,* Finite Rings with Identity (1974)
29. *J. Satake,* Linear Algebra (S. Koh, T. Akiba, and S. Ihara, translators) (1975)

30. *J. S. Golan*, Localization of Noncommutative Rings (1975)
31. *G. Klambauer*, Mathematical Analysis (1975)
32. *M. K. Agoston*, Algebraic Topology: A First Course (1976)
33. *K. R. Goodearl*, Ring Theory: Nonsingular Rings and Modules (1976)
34. *L. E. Mansfield*, Linear Algebra with Geometric Applications: Selected Topics (1976)
35. *N. J. Pullman*, Matrix Theory and Its Applications (1976)
36. *B. R. McDonald*, Geometric Algebra Over Local Rings (1976)
37. *C. W. Groetsch*, Generalized Inverses of Linear Operators: Representation and Approximation (1977)
38. *J. E. Kuczkowski and J. L. Gersting*, Abstract Algebra: A First Look (1977)
39. *C. O. Christenson and W. L. Voxman*, Aspects of Topology (1977)
40. *M. Nagata*, Field Theory (1977)
41. *R. L. Long*, Algebraic Number Theory (1977)
42. *W. F. Pfeffer*, Integrals and Measures (1977)
43. *R. L. Wheeden and A. Zygmund*, Measure and Integral: An Introduction to Real Analysis (1977)
44. *J. H. Curtiss*, Introduction to Functions of a Complex Variable (1978)
45. *K. Hrbacek and T. Jech*, Introduction to Set Theory (1978)
46. *W. S. Massey*, Homology and Cohomology Theory (1978)
47. *M. Marcus*, Introduction to Modern Algebra (1978)
48. *E. C. Young*, Vector and Tensor Analysis (1978)
49. *S. B. Nadler, Jr.*, Hyperspaces of Sets (1978)
50. *S. K. Sehgal*, Topics in Group Rings (1978)
51. *A. C. M. van Rooij*, Non-Archimedean Functional Analysis (1978)
52. *L. Corwin and R. Szczarba*, Calculus in Vector Spaces (1979)
53. *C. Sadosky*, Interpolation of Operators and Singular Integrals: An Introduction to Harmonic Analysis (1979)
54. *J. Cronin*, Differential Equations: Introduction and Quantitative Theory (1980)
55. *C. W. Groetsch*, Elements of Applicable Functional Analysis (1980)
56. *I. Vaisman*, Foundations of Three-Dimensional Euclidean Geometry (1980)
57. *H. I. Freedman*, Deterministic Mathematical Models in Population Ecology (1980)
58. *S. B. Chae*, Lebesgue Integration (1980)
59. *C. S. Rees, S. M. Shah, and Č. V. Stanojević*, Theory and Applications of Fourier Analysis (1981)
60. *L. Nachbin*, Introduction to Functional Analysis: Banach Spaces and Differential Calculus (R. M. Aron, translator) (1981)
61. *G. Orzech and M. Orzech*, Plane Algebraic Curves: An Introduction Via Valuations (1981)
62. *R. Johnsonbaugh and W. E. Pfaffenberger*, Foundations of Mathematical Analysis (1981)

63. *W. L. Voxman and R. H. Goetschel,* Advanced Calculus: An Introduction to Modern Analysis (1981)
64. *L. J. Corwin and R. H. Szczarba,* Multivariable Calculus (1982)
65. *V. I. Istrăţescu,* Introduction to Linear Operator Theory (1981)
66. *R. D. Järvinen,* Finite and Infinite Dimensional Linear Spaces: A Comparative Study in Algebraic and Analytic Settings (1981)
67. *J. K. Beem and P. E. Ehrlich,* Global Lorentzian Geometry (1981)
68. *D. L. Armacost,* The Structure of Locally Compact Abelian Groups (1981)
69. *J. W. Brewer and M. K. Smith, eds.,* Emmy Noether: A Tribute to Her and Work (1981)
70. *K. H. Kim,* Boolean Matrix Theory and Applications (1982)

Other Volumes in Preparation

BOOLEAN MATRIX THEORY AND APPLICATIONS

Ki Hang Kim

Mathematical Research Group
Alabama State University
Montgomery, Alabama

MARCEL DEKKER, INC. New York and Basel

Library of Congress Cataloging in Publication Data

Kim, Ki Hang.
  Boolean matrix theory and applications.

  (Monographs and textbooks in pure and applied mathematics ; 70)
  Includes index.
  1. Algebra, Boolean. 2. Matrices. I. Title.
II. Series.
QA10.3.K57   511.3'24   82-1481
ISBN 0-8247-1788-0   AACR2

MARCEL DEKKER, INC.
270 Madison Avenue, New York, New York 10016

Current printing (last digit):
10 9 8 7 6 5 4 3 2 1

PRINTED IN THE UNITED STATES OF AMERICA

In memory of my father

Jin Gyong Kim

FOREWORD

Lucus a non lucendo, so goes the Latin saying. Once meant as a preposterous sample of miscarried etymology, we can now take this proverb as a suitable slogan for a recurrent phenomenon in the development of mathematics. Borrowing Hans Christian Andersen's insight, we could also call it "the ugly duckling phenomenon."

When E. H. Moore, the genius cursed with a crackpot notation, came out with his classification of finite fields, no one would have bet a penny that these monstrous gnomes, long the bane of analysts fulfilling their graduate requirement in algebra, would ever be taken seriously by the engineers at A.T.&T. And yet, nowadays billions are saved by the communications industry by employing codes on finite fields.

And what about logic? Only a generation ago, a mathematical logician would be taken either as a philosopher with misguided mathematical ambitions, or as a mathematician with a closet philosophical orientation, depending on which of the two departments the person happened to belong to. That is, until the day came when model theory began to make dangerous inroads into algebra; when unexpected corners of mathematics began to be threatened by proof theory, or Centzenizatic as the threat was maliciously dubbed by someone who was thinking of Finland; or when recursion theory became the blueprint for much work in computer science.

Another case in point is category theory. The unabashed success of the insightful and liberating definitions in this field is a hard pill to swallow by the mathematician who is still wedded to the view of mathematics consisting of a succession of hard-won pearls called theorems, which he then reluctantly tosses out to a public intent on generalizing rather than understanding.

In the sleepy days when the provinces of France were still quietly provincial, matrices with Boolean entries were a favored occupation of aging professors at the universities of Bordeaux and Clermont-Ferrand. But one day, extremal set theory came along, and caught the fancy of some of the best combinatorial minds of our time: Erdős, Graham, Katona, Kleitman, Lovasz, Rado, and Spencer, to name a few. By all reckoning, extremal set theory is now rivaling number theory in style as well as in depth, with the added boon of having substantial applications. Who would have thought that Boolean matrices would turn out to be a suitable notation and techniques for the disparate problems of extremal set theory ? Professor Kim, in this lucid and exceptionally thorough book, makes a strong case for this approach. He has left no stone unturned. The reader will find here the fulfillment of a rare event: the complete exposition of all that has been done in a field, both in the theory and in the applications. The subject is far from reaching maturity, and Kim's book will allow the student, for once, to learn all that has been done so far without being irritatingly referred elsewhere. We owe Kim our thanks for his efforts, and for setting an impeccable standard of clarity and logical organization.

Gian-Carlo Rota
Massachusetts Institute of Technology
Cambridge, Massachusetts

PREFACE

The purpose of this book is to present algebraic and combinatorial aspects of Boolean matrices and their applications. This is a first attempt to survey Boolean matrices and the theory related to them. Most of the results have appeared in various journals and other monographs in the midst of other topics. Since the subject is so vast, I have selected from it material close to my own work. I have almost completely ignored applications to computer science because of the huge amount of material in this area. It alone could be the subject for a second volume. The manuscript is an outgrowth of lectures given at the Instituto de Fisica e Matematica, Lisbon, Portugal; the University Of Delhi, Delhi, India; the University of Stuttgart, Stuttgart, Germany; and at various other institutions over a period of ten years.

Applications of the theory of Boolean matrices are of fundamental importance in the formulation and analysis of many classes of discrete structural models which arise in the physical, biological, and social sciences. The theory is also intimately related to many branches of mathematics, including relation theory, logic, graph theory, lattice theory, and algebraic semigroup theory. Since these subjects are well-developed, we shall choose techniques and viewpoints suitable to them, and an approach which is useful for these applications. However we will not present results from these areas unless they have direct bearing on our theory and applications.

An understanding of the material presented here should allow the reader to continue his own study in more general aspects of Boolean matrices. In addition he should be able to appreciate that applications of these matrices include many subjects, and many further applications seem likely.

A Boolean matrix is a matrix over a Boolean algebra. These matrices require a completely separate treatment from that of matrices over the real field $\mathcal{R}$ or complex field $\mathcal{C}$ due to the fact that Boolean matrices do not satisfy many of the fundamental properties of matrices over $\mathcal{R}$ or $\mathcal{C}$. For example, a Boolean matrix need not always possess a g-inverse, in sharp contrast to the real (complex) cases where every matrix has a g-inverse. Also, for Boolean matrices, row and column rank need not be equal, which is again a fundamental result for matrices over $\mathcal{R}$ and $\mathcal{C}$. Another contradiction to our basic concept for real and complex field is that a set of k-tuples over the Boolean algebra may have more than k independent vectors. Due to these interesting deviations from the general complex field, a few of the usual definitions for the real (complex) field need modifications to be meaningful for Boolean algebra.

In Chapter 1 we give definitions regarding Boolean vectors and Boolean matrices. Properties of sets of Boolean vectors such as dependence and independence and existence of bases are discussed. We define and study three concepts of rank of a Boolean matrix: row rank, column rank, and Schein rank, and prove the elementary characterizations of Green's relations for Boolean matrices. Throughout the book some advanced material (which the reader may skip if he wishes) is interspersed with more fundamental results. Chapter 1 also contains material on applications to diagnosis of disease, symbolic logic, circuit design, size of R-classes, asymptotic numbers of D, L, R-classes, combinatorial set theory, eigenvectors, and square roots of Boolean matrices.

In Chapter 2, the more fundamental results are concerned with the study of regular matrices and idempotents.

In Chapter 3 we discuss different generalized inverses of Boolean matrices. Then we discuss algorithms for obtaining these inverses. Most of the results are fairly elementary but are not needed for later chapters.

In Chapter 4 we discuss combinatorial properties of order relations. There are one-to-one correspondences between isomorphism classes of finite topologies, quasi-orders, and regular D-classes. Much effort has been devoted to the still unsolved problem of enumerating all finite topologies on a set of n elements. Two special classes of binary relations with a number of applications are similarity relations and connective relations. Order relations have important applications to economics, to theorems of Arrow, and others about social welfare functions.

Most of Chapter 5 is concerned with questions such as do large powers of a Boolean matrix converge to a constant matrix? This can be studied by graphical methods, or analytically. If large powers converge to a matrix having all ones (which is frequently the case), the matrix is called primitive. Special classes of primitive matrices such as fully indecomposable and nearly decomposable Boolean matrices are important. The most celebrated theorem in the theory of the powers of a Boolean matrix of order n asserts that all powers after the $(n - 1)^2$ power are periodic, but many generalizations and other results have been obtained. We then survey applications to sociology, finite automata, parallel computation, information systems, analysis of switching circuits, and give references to other applications.

A brief appendix shows how many results about Boolean matrices over $\{0, 1\}$ extend to matrices over an arbitrary Boolean algebra.

Following this a list of problems which are unsolved as far as we know, is given. Some are extremely difficult. Perhaps many might be good topics for further research.

The exercises should help the reader to review, assimilate, and increase his skill in the material in each part of the book. They vary widely in difficulty: some are computations, others involve reasoning which can be used in many situations, and some involve difficult proofs which might challenge a professional mathematician. The reader can select among them according to his purpose in reading the book. The material covered by the exercises is in approximately the same order in which it occurs in the book.

In the preface to S. Rudeanu [2], the late Professor G. Moisil indicated that a single chapter of that book sketched some elements of a theory of Boolean matrices, the full development of which is left to the future. The present book is a continuation of this line of thought, and perhaps a partial fulfillment of this remark.

Ki Hang Kim

ACKNOWLEDGMENTS

In preparing this book, I have benefited from the assistance of many people. Especially, I owe considerable thanks to my colleague and friend, Professor Fred W. Roush, who read the manuscript at each stage of its development and whose criticisms and suggestions have resulted in major improvements in both content and style. I am grateful to Professors Jozsef Denes, Joe Neggers, Gian-Carlo Rota, and Boris M. Schein for kindly reading the earlier drafts and making very useful comments and suggestions. In addition Professor Boris M. Schein has been of valuable assistance in supplying Russian references. The author is also indebted to Prof. Thomas S. Blyth, Dr. Gyula Katona, Prof. James R. Krabill, Prof. R. Duncan Luce, Dr. George Markowsky, Prof. Graciano de Oliveira, Dr. David Rosenblatt, Prof. E. Fried, Dr. F. Pintz, Prof. R. P. Sullivan, and Dr. M. Szalay for reading part of the rough draft and providing very useful criticism. However, I alone am responsible for any remaining errors and most of the off-key remarks. Thanks are also due to Mrs. Kay Roush, who proofread the manuscript with her usual accuracy and diligence.

CONTENTS

BOOLEAN MATRIX THEORY AND APPLICATIONS

CHAPTER ONE

FUNDAMENTAL CONCEPTS

This chapter is devoted to explaining the fundamental concepts of Boolean vectors and Boolean matrices.

In order to develop the theory of Boolean matrices we must begin with the concept of Boolean algebra.

The definition of a Boolean algebra which we are about to present is based on a structure introduced by E. V. Huntington in 1904.

A *Boolean algebra* is a mathematical system $(\beta, +, \cdot)$ consisting of a nonempty set β and two binary operations + and $\cdot$ defined on β such that (1) each of the operations + and $\cdot$ is commutative, that is, $a + b = b + a$, $a \cdot b = b \cdot a$ for all $a, b \in \beta$; (2) each operation is distributive over the other, that is, $a + (b \cdot c) = (a + b) \cdot (a + c)$, $a \cdot (b + c) = (a \cdot b) + (a \cdot c)$ for all $a, b, c \in \beta$; (3) there exist distinct identity element 0 and 1 relative to the operations + and $\cdot$, respectively, that is, $a + 0 = a$, $a \cdot 1 = a$ for all $a \in \beta$; (4) for each element $a \in \beta$, there exists an element $a^c \in \beta$, called the *complement* of a, such that $a + a^c = 1$, $a \cdot a^c = 0$.

Boolean algebra is named after the British mathematician George Boole (1813-1864). For elementary properties of Boolean algebra, the reader is referred to S. Rudeanu [327], F. E. Hohn [155], and A. Abian [1]. For an application of Boolean algebra for operations research, see P. L. Hammer and S. Rudeanu [136].

The Boolean algebra of two elements is most frequently used in applications, and all other finite Boolean algebras are simply direct sums of copies of it. Moreover in the Appendix we show that by homomorphisms the theory of matrices over any Boolean algebra reduces to the two element case. Thus we will primarily work with the two element Boolean algebra in this book. We shall use β_0 to denote the set $\{0, 1\}$ with three operations $+, \cdot, {}^c$ defined as follows: $0 + 0 = 0 \cdot 1 = 1 \cdot 0 = 0 \cdot 0 = 0$, $1 + 0 = 0 + 1 = 1 + 1 = 1 \cdot 1 = 1$, $0^c = 1$, and $1^c = 0$. We usually suppress the dot "$\cdot$" of $a \cdot b$ and simply write ab. Hereafter, by a vector or matrix we shall mean a vector or matrix over β_0 unless stated otherwise. Matrices will be denoted by capital letters A, B, etc., and vectors by lower case letters.

Its best known use is in symbolic logic, where 0 is "false," 1 is "true," and the operations $+, \cdot, {}^c$ are "and," "or," "not.". However it also has other interpretations: in switching circuits 1, 0 could mean current is, or is not flowing and $+, \cdot$ could refer to switches being connected in parallel or series. Another important interpretation is that 0 can refer to the number 0 and 1 to the class of positive numbers. Thus $1 + 1 = 1$ means "a positive number plus a positive number is positive." Moreover, the 0 and 1 reflects the philosophical thought embodied in symbols of dualism of the cosmos in *I Ching* (The Classic of Changes or The Book of Divinition) of ancient China. That is, 0 represents *Yin* (negative force, earth, bad, passive, destruction, night, autumn, short, water, cold, etc.) and 1 represents *Yang* (positive force, heaven, good, active, construction, day, spring, tall, fire, hot, etc.).

A *semigroup* is a set closed under an associative binary operation. We recognize, of course, that β_0 is commutative semigroup with respect to + and $\cdot$. In fact, it is easy to see that both distributive laws hold in β_0.

A *semiring* is an additive and multiplicative system which is closed and associative under both operations, and commutative under addition, satisfying also the distributive law. Thus a system $(\beta_0, +, \cdot)$ forms a semiring.

By a *Boolean matrix* we mean matrix over β_0. Such a matrix can be interpreted as a binary relation. There are many papers, books, lecture notes and, dissertations devoted to binary relations and it is a well-shaped and vital part of mathematics [399, 341, 153, 228, 249, 250, 268].

The (binary) relation has a long history. Relation, as a subject matter for logical investigation, was mentioned by Aristotle, and is among his categories. But he did little more than mention it [59]. The modern treatment of relations began with De Morgan, but was first adequately developed by C. S. Peirce [276]. Peirce's work was taken over and systematically extended by E. Schröder [341]. The best known presentation of the logic of relation is that of A. N. Whitehead and B. Russell [399] in their *Principia Mathematica*.

There exist many papers concerning the abstract characterization of the semigroup of relations and its subsemigroups. There are also many papers dealing with semigroups which may be represented as a semigroup of relations. Among the general references to studies on the subject we would mention S. Schwarz [345, 347], B. M. Schein [337, 338], R. J. Plemmons and M. T. West [286], A. D. Wallace [395], J. Riguet [311], K. A. Zaretski [409, 410], L. M. Gluskin [124], J. S. Montague and R. J. Plemmons [257], K. H. Kim [170, 180], G. Markowsky [244, 245], and R. J. Simpson [369].

However, relatively little work has been done on the Boolean matrices. Matrix notation was applied to the logic of relations as long ago as 1970 by C. S. Peirce [276], who was at that time unaware of Cayley's work [59]. He seems to have been led to introduce matrices partly as an aid in his classfication of relations and partly for the sake of illustrations or examples. C. Berge [21], F. Harary [137], F. Harary, R. Z. Norman, and D. Cartwright [140], and some others have each devoted a chapter or two to Boolean matrices as a computational method in graph theory and network theory. Virtually all of the recent work has applications to mathematics and other scientific disciplines, e.g., Perron-Frobenius theory of nonnegative matrices, Markov chains, etc. [321].

We will use standard-set-theoretical notation. In this book we will be dealing with the following sets:

(1) The set of positive integers Z^+.

(2) The set of all integers Z.

(3) The field of rational numbers Q.

(4) The field of real numbers R.

(5) The field of complex numbers C.

Let $\underline{n} = \{1, 2, \ldots, n\}$. Let $|\underline{n}|$ denote the cardinality of $\underline{n}$. On occasion, when it is clear what set the index ranges over we will simplify expressions, e.g., instead of

$$\sum_{i=0}^{k} a_i$$

we might simply write

$$\sum_{\underline{m}} a_i$$

where $\underline{m} \subset \underline{n}$ or even Σa_i.

1.1 Boolean Vectors

In order of development, the next structure we take up is Boolean vector. A Boolean vector is an n-tuple of elements from a Boolean algebra. Such n-tuples can be added, multiplied by elements of the Boolean algebra, or operated upon by Boolean matrices. Although it is their relationship to Boolean matrices which is of prime importance in this book, Boolean vectors and Boolean vector spaces are of some independent interest in combinatorics. For instance we can associate a Boolean vector to any subset of a set.

Most of the material in this section may be found in K. H. Kim [178] and G. Markowsky [244].

DEFINITION 1.1.1. Let V_n denote the set of all n-tuples $(a_1, a_2, \ldots, a_n)$ over β_0. An element of V_n is called a *Boolean vector* of dimension n. The system V_n together with the operation of component-wise addition is called the *Boolean vector space of dimension n.*

Note that V_n can be made into a Boolean algebra by defining two operations +, · on it as follows: $(a_1, a_2, \ldots, a_n) + (b_1, b_2, \ldots, b_n) = (a_1 + b_1, a_2 + b_2, \ldots, a_n + b_n)$ and $(a_1, a_2, \ldots, a_n)(b_1, b_2, \ldots, b_n) = (a_1b_1, a_2b_2, \ldots, a_nb_n)$ for all $a_i, b_i \in \beta_0$.

DEFINITION 1.1.2. Let $V^n = \{v^T: v \in V_n\}$, where v^T we mean *column vector*

$$\begin{bmatrix} v_1 \\ v_2 \\ \vdots \\ v_n \end{bmatrix}$$

DEFINITION 1.1.3. Define the *complement* v^C of v to be the vector such that $v_i^C = 1$ if and only if $v_i = 0$, where v_i denote the ith component of v.

Let $u, v \in V^n$, then we define $uv = (u^Tv^T)$, $u + v = (u^T + v^T)^T$, $(u^C)^T = (u^T)^C$ and $u_k = u_k^T$. Thus V^n is isomorphic to V_n as a Boolean algebra. And so we will usually discuss V_n.

DEFINITION 1.1.4. Let e_i be the n-tuple with 1 as ith coordinate, 0 otherwise. Further, we define $e^i = e_i^T$.

We will usually write 0 instead of (0, 0, ... , 0) or $(0, 0, \ldots, 0)^T$ since usually it will be clear from the context what we mean.

DEFINITION 1.1.5. A *subspace* of V_n is a subset containing the zero vector and closed under addition of vectors. The *span* of a set W of vectors, denoted <W>, is the intersection of all subspaces containing W.

Note that $<W>$ is the set of all finite sums of elements of W, providing that we accept the convention that the empty sum is equal to 0.

EXAMPLE 1.1.1. (1) Let $W_1 = \{(0\ 0\ 0), (1\ 0\ 0), (0\ 1\ 0), (0\ 0\ 1), (1\ 1\ 0), (0\ 1\ 1), (1\ 0\ 1), (1\ 1\ 1)\}$. Then W_1 is a subspace of V_3.

(2) Let $W_2 = \{(0\ 0\ 0), (1\ 0\ 0), (0\ 0\ 1), (1\ 1\ 0), (0\ 1\ 1), (1\ 1\ 1)\}$. Then W_2 is not a subspace of V_3.

DEFINITION 1.1.6. Let $W \subset V_n$. A vector $v \in V_n$ is said to be *dependent* on W if and only if $v \in <W>$. A set W is said to be *independent* if and only if for all $v \in W$, v is not dependent on $W \setminus \{v\}$, where $\setminus$ denotes the set-theoretic difference. If W is not independent we say that it is dependent.

DEFINITION 1.1.7. Let $u, v \in V_n$, then we say that $u \leq v$ if and only if $v_i = 1$ for each i such that $u_i = 1$. We say that $u < v$ if $u \leq v$ and $u \neq v$.

DEFINITION 1.1.8. Let $W \subset V_n$. A subset B of W is called a *basis* of W if and only if $W = <B>$, and B is an independent set. (Note that the basis of 0 is $\emptyset$).

Observe that $\{e_1, e_2, \ldots, e_n\}$ is the basis of V_n, and that if B is a basis of a subspace W, then every element of W is a finite sum of elements of B. Note also that all elements of V_n are idempotent with respect to addition and multiplication, so that there are only finitely many distinct sums possible. Moreover, if W is a subspace then $W = <W>$.

THEOREM 1.1.1 [178]. Let W be a subspace of V_n. Then there exists one subset B of V_n such that B is a basis of W.

Proof. Let B be the set of all vectors of W that are not sums of vectors of W smaller than themselves. Then B is an independent set. Suppose B generates a proper subspace of W. Then let v be a minimal vector of $W \setminus \langle B \rangle$. Then v is expressible as a sum of smaller vectors since it is not in B. But these smaller vectors must be in $\langle B \rangle$ since v was minimal. Thus v belongs in the subspace generated by B. This is a contradiction. Thus B generates W and is a basis. By independence, B must be contained in every basis. Let B' be another basis and let u be a minimal element of $B' \setminus B$. Then by the reasoning above u is dependent. This contradiction shows B' = B. This proves the theorem. □

1.2 Boolean Matrices

Here we shall say what a Boolean matrix is and how Boolean matrices are added and multiplied. (The method is really the same as for matrices of complex numbers except that addition and multiplication of individual entries is Boolean). It will develop in the course of the book that Boolean matrices have quite different properties from matrices over a field, due to the fact that addition in a Boolean algebra does not form a group. Every Boolean linear transformation on V_n can be represented by a unique Boolean matrix. Boolean matrices may arise from graphs or from nonnegative real matrices by replacing all positive entries by 1, but their most frequent occurrence is in the representation of binary relations. At the end of this section we will include three specific applications.

One of the most important ways to study a Boolean matrix is to consider its row space, that is, the subspace of V_n spanned by its rows. The row space forms a special type of partially ordered set called a lattice. Conversely every lattice can be conveniently represented as the row space of some Boolean matrix.

We shall illustrate the basic facts of Boolean matrices. This section uses the work of K. H. Kim [178] and G. Markowsky [244].

DEFINITION 1.2.1. By a *Boolean matrix* of size $m \times n$ is meant an $m \times n$ matrix over β_0. Let B_{mn} denote the set of all $m \times n$ such matrices. If $m = n$, we write B_n. Elements of B_{mn} are often called *relation matrices*, *Boolean relation matrices*, *binary relation matrices*, *binary Boolean matrices*, (0, 1)-*Boolean matrices*, and (0, 1)-*matrices*.

DEFINITION 1.2.2. Let $A = (a_{ij}) \in B_{mn}$. Then the element a_{ij} is called the (i, j)-*entry* of A. The (i, j)-entry of A is sometimes designated by A_{ij}. The ith *row* of A is the sequence $a_{i1}, a_{i2}, \ldots, a_{in}$, and jth *column* of A is the sequence $a_{1j}, a_{2j}, \ldots, a_{mj}$. Let A_{i*} (A_{*i}) denote the ith row (column) of A.

Note that a row (column) vector is just an element of B_{1n} (B_{m1}).

DEFINITION 1.2.3. The $n \times m$ *zero matrix* 0 is the matrix all of whose entries are zero. The $n \times n$ *identity matrix* I is the matrix (δ_{ij}) such that $\delta_{ij} = 1$ if $i = j$ and $\delta_{ij} = 0$ if $i \neq j$. The $n \times m$ *universal matrix* J is the matrix all of whose entries are 1.

Since the order of the matrix is clear from the context, most of the time we suppress the order of the matrix. Matrix addition and multiplication are the same as in the case of complex matrices but the concerned sums and products of elements are Boolean. Concepts such as transpose, symmetricity, and idempotency, etc., are the same as in the case of real or complex matrices.

EXAMPLE 1.2.1. If

$$A = \begin{bmatrix} 1 & 1 & 1 \\ 1 & 0 & 1 \\ 0 & 1 & 0 \end{bmatrix}, \qquad B = \begin{bmatrix} 1 & 1 & 1 \\ 1 & 1 & 1 \\ 0 & 1 & 1 \end{bmatrix}$$

then

$$A + B = \begin{bmatrix} 1 & 1 & 1 \\ 1 & 1 & 1 \\ 0 & 1 & 1 \end{bmatrix}, \qquad AB = \begin{bmatrix} 1 & 1 & 1 \\ 1 & 1 & 1 \\ 1 & 1 & 1 \end{bmatrix}$$

DEFINITION 1.2.4. We use the notation A^2 to designate the product AA, $A^3 = AA^2$, and in general $A^k = AA^{k-1}$ for any positive integer k. The matrix A^k is called the kth *power* for obvious reasons. The notations $a_{ij}^{(k)}$ and $A_{ij}^{(k)}$ mean (i, j)-entry and (i, j)-block of A^k. The notation $(A_{ij})^k$ means the kth power of the (i, j)-block of A.

DEFINITION 1.2.5. A *binary relation* on a set X is a subset of X × X. The *composition* of two binary relations ρ_1, ρ_2 is the relation γ such that $(x, y) \in \gamma$ if and only if for some z both $(x, z) \in \rho_1$ and $(z, y) \in \rho_2$.

The matrix of the composition of two binary relations will be the Boolean product of the matrices of the two relations. Furthermore, we can interpret a Boolean matrix as a graph.

DEFINITION 1.2.6. The *adjacency matrix* A_G of a directed graph (digraph) G is the matrix $A_G = (a_{ij})$ such that $a_{ij} = 1$ if there is an arc from vertex v_i to vertex v_j and $a_{ij} = 0$ otherwise. Dually, a digraph G determines and is determined by the Boolean matrix A_G.

The usual product G_1G_2 of digraphs with vertex set has as its adjacency matrix $A_{G_1}A_{G_2}$. Because of these correspondences, it is natural to employ the terminology and notation of both the binary relation theory and the graph theory to solve problems in the Boolean matrix theory. For the connection between digraphs and Boolean matrices, see C. Berge [21], F. Harary, R. Z. Norman, and D. Cartwright [140]. However, for a discussion of relationship between binary relations and digraphs, see F. Harary [137], S. S. Anderson [9], and O. Ore [271]. It is recommended that the reader interested in digraphs consult F. Harary, R. Z. Norman, and D. Cartwright [140].

EXAMPLE 1.2.2. Let

$$A = \begin{bmatrix} 1 & 0 & 0 \\ 1 & 1 & 1 \\ 0 & 0 & 1 \end{bmatrix}$$

Then the binary relation for A is $\{(x_1, x_1), (x_2, x_1), (x_2, x_2), (x_2, x_3), (x_3, x_3)\}$, and the digraph for A is

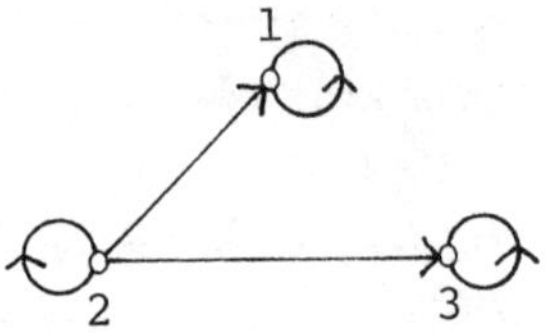

For the connection between (0, 1)-matrices over $\mathcal{R}$ and digraphs, the reader is referred to A. L. Dulmage and N. S. Mendelsohn [95], F. Harary [137], S. S. Anderson [9], C. Berge [21], and O. Ore [271]. For basic properties of these matrices, see M. Marcus and H. Minc [241], and H. J. Ryser [331].

It is easy to see that B_n forms a semigroup under multiplication. However, B_{mn} does not form a semigroup under multiplication. For various algebraic properties of semigroups, the reader should consult A. H. Clifford and G. B. Preston [55], P. Dubreil [94], E. S. Ljapin [229], M. Petrich [280, 278], L. Redei [299], G. Pappy [272], K. M. Kapp and H. Schneider [169], K. Krohn, J. Rhodes, and B. Tilson [223], and A. Suskevitch [379].

DEFINITION 1.2.7. A square matrix is called a *permutation matrix* if every row and every column contains exactly one 1. Let P_n denote the set of all $n \times n$ such matrices.

DEFINITION 1.2.8. The *transpose* of a Boolean matrix is obtained by rewriting its rows and columns. The transpose of A will be denoted by A^T.

Note that if $A \in P_n$, then $AA^T = A^TA = I$. Moreover, the set P_n forms a subgroup of the semigroup B_n.

DEFINITION 1.2.9. A matrix is said to be a *partial permutation matrix* if every row and every column of it contains at most one 1.

DEFINITION 1.2.10. Let $A, B \in B_{mn}$. By $B \le A$ we mean if $b_{ij} = 1$ then $a_{ij} = 1$ for every i and j.

We remark that if A is a partial permutation matrix then $AA^T \le I$ and $A^TA \le I$.

DEFINITION 1.2.11. The *row space* of a matrix A is the span of the set of all rows of A. Likewise one has a *column space*. Let R(A) (C(A)) denote the row (column) space of A.

EXAMPLE 1.2.3. Let

$$A = \begin{bmatrix} 1 & 0 & 0 \\ 1 & 1 & 0 \\ 1 & 0 & 1 \end{bmatrix}$$

Then R(A) = {(0 0 0), (1 0 0), (1 1 0), (1 0 1), (1 1 1)}, and C(A) = $\{(0\ 0\ 0)^T, (0\ 1\ 0)^T, (0\ 0\ 1)^T, (0\ 1\ 1)^T, (1\ 1\ 1)^T\}$.

DEFINITION 1.2.12. By a *partial order relation* on a set X we mean a reflexive, antisymmetric and transitive relation on X. A set X together with a specific partial order relation P in X is called a partially ordered set (poset). A *linear order* (also called *total order*) is a partial order relation P such that for all x, y; $(x, y) \in P$ or $(y, x) \in P$.

DEFINITION 1.2.13. A *lattice* is a partially ordered set in which every pair of elements has a least upper bound (join) and greatest lower bound (meet). The operations join and meet are denoted by $\vee$ and $\wedge$ respectively.

EXAMPLE 1.2.4. Examples are the set of real numbers under $\leq$, the family of all subsets of a set under $\subseteq$, and the families of open or closed sets of any topology under $\subseteq$.

DEFINITION 1.2.14. A lattice is said to be *distributive* if and only if $A \vee (B \wedge C) = (A \vee B) \wedge (A \vee C)$ for all A, B, C. This is equivalent to the dual condition $A \wedge (B \vee C) = (A \wedge B) \vee (A \wedge C)$.

EXAMPLE 1.2.5. The lattice of all sets is distributive, as are the lattices of open sets of any topology, since in these cases $\wedge$ is $\cap$ and $\vee$ is $\cup$.

For elementary properties of lattice theory, the reader should consult Chapter 1 of G. Birkhoff [22].

If $A \in B_{mn}$, then R(A) (C(A)) is a lattice. Then the join of two elements is their sum, while the meet of two elements is the sum of the elements of R(A) which are less than or equal to both elements. Of course 0 is the universal lower bound while the sum of all the elements of R(A) is the universal lower bound. To add clarity to the exposition we will often write La R(A) (La C(A)) instead of simply R(A) (C(A)) whenever we wish to call attention to the fact that R(A) (C(A)) is a lattice.

Until recently, only a very few authors have attempted to point out the close relationship that exists between the lattices and binary relations. However, discussion of such relationships may be found in K. A. Zaretski [412] and G. Markowsky [243, 244, 245].

PROPOSITION 1.2.1. If $A \in B_{mn}$, then R(A) (C(A)) is a subspace of V_n (V^m).

Proof. This relation holds for matrices over any semiring, and the proof is the same as in the case of matrices over $\mathcal{R}$. $\square$

EXAMPLE 1.2.6. Let A be as in Example 1.2.3.

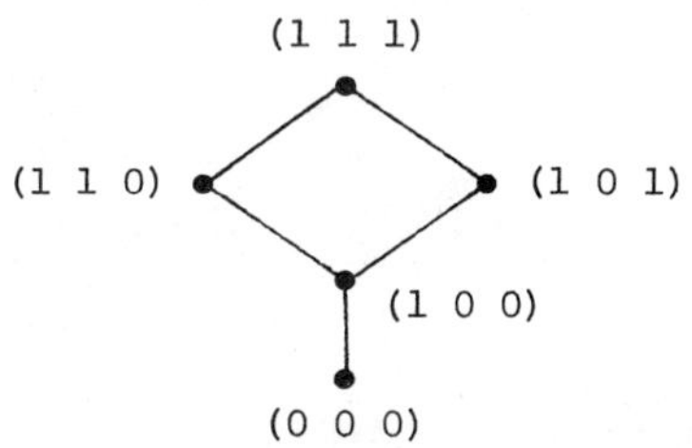

Ra R(A)

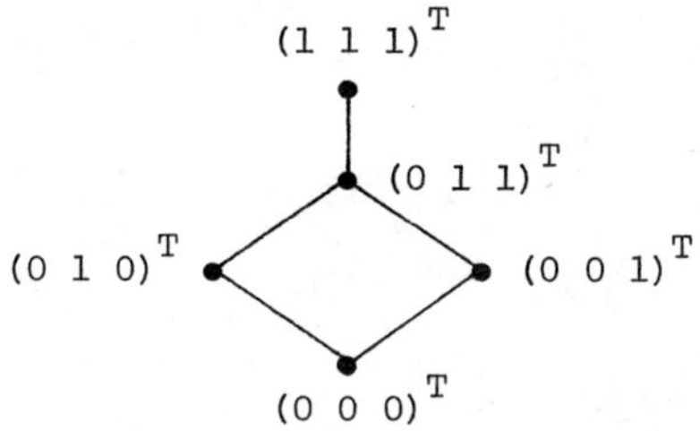

La C(A)

PROPOSITION 1.2.2. Let $A \in B_{mk}$, $B \in B_{kn}$, then $R(AB) \subseteq R(B)$ and $C(AB) \subseteq C(A)$.

Proof. The proof follows immediately from the fact that the rows of AB are sums of the rows of B, etc. □

THEOREM 1.2.3 [244]. If $A \in B_{mn}$, then $|C(A)| = |R(A)|$.

Proof. We shall construct a bijection between C(A) and R(A). Let $v \in C(A)$, then there exists a unique set $\underline{s} \subset \underline{m} \subseteq \underline{n}$ such that

$$v = \sum_{\underline{s}} e^i$$

Let $\underline{s}' = \underline{m} \setminus \underline{s}$ (we will use ' to denote set complements in this proof). Consider the map $f\colon C(A) \to R(A)$ given by

$$f(v) = \sum_{\underline{s}'} A_{i*}$$

where $v \in C(A)$. Clearly f is well-defined. We claim that the following statements are true.

(1) f is injective: suppose we have $v, w \in C(A)$, $v \neq w$, and

$$v = \sum_{\underline{s}} e^i$$

while

$$w = \sum_{\underline{t}} e^i$$

where $\underline{t} \subset \underline{m}$ and that $f(v) = f(w)$, i.e.,

$$\sum_{\underline{s}'} A_{i*} = \sum_{\underline{t}'} A_{i*}$$

Since $v \neq w$, we may assume that there exists a $p \in \underline{t} \setminus \underline{s}$. But since $w \in C(A)$ there exists a $k \in \underline{n}$ such that $a_{pk} = 1$ and $A_{*k} \leq w$. Since $p \in \underline{s}'$, we must have $(f(v))_k = 1$, which implies that there exists a $q \in \underline{t}'$ such that $a_{qk} = 1$. But since $A_{*k} \leq w$, this implies that $e^q \leq w$, which is impossible since $q \in t'$. Thus $f(v) \neq f(w)$.

(2) Since there exists an injection from the row space into the column space, we are through as it follows that f is surjective. □

COROLLARY 1.2.4 [244]. Let A and f be as in Theorem 1.2.3, and let $v, w \in C(A)$. Then $v \leq w$ if and only if $f(v) \geq f(w)$.

Proof. Necessity. Clear, in that $\underline{s} \subset \underline{t}$ if and only if $\underline{s}' \supset \underline{t}'$. Sufficiency. This follows from the same proof as that given for injectivity of f. Assume $w \not\leq v$ but $f(w) \geq f(v)$, and follow exactly the same procedure. □

In view of the preceding corollary, without loss of generality we may assume that f is an anti-isomorphism of lattices from R(A) onto C(A).

PROPOSITION 1.2.5 [244]. Let $A_1, A_2, \ldots, A_k \in B_n$ and let $B = A_1A_2 \ldots A_k$. Then $|C(B)| = |R(B)| \leq |R(A_i)| = |C(A_i)|$ for all i.

Proof. For any M, N we have $|R(MN)| \leq |R(N)|$ and $|R(MN)| = |C(MN)| \leq |C(M)| = |R(M)|$, by Proposition 1.2.2. The present proposition follows from this by induction. □

DEFINITION 1.2.15. Let $A \in B_{mn}$. By $B_r(A)$ we mean the unique basis of $R(A)$, and we call it the *row basis* of A. Similarly, by $B_c(A)$ we mean the unique basis of $C(A)$, which we call the *column basis* of A. The cardinality of $B_r(A)$ ($B_c(A)$) is called the *row* (*column*) *rank* of A and is denoted by $\rho_r(A)$ ($\rho_c(A)$).

Note that $\rho_r(0) = \rho_c(0) = 0$. In general, $\rho_r(A) \neq \rho_c(A)$ for $n > 3$. It should be pointed out that the row and column rank of a matrix are unaltered by premultiplying and postmultiplying by permutation matrices. This is known as *transformation*, or *conjugation*, of a matrix.

EXAMPLE 1.2.7. If

$$A = \begin{bmatrix} 0 & 1 & 0 \\ 0 & 0 & 1 \\ 1 & 0 & 1 \end{bmatrix}$$

then $B_r(A) = \{(0\ 1\ 0), (0\ 0\ 1), (1\ 0\ 1)\}$, $B_c(A) = \{(0\ 0\ 1)^T, (1\ 0\ 0)^T, (0\ 1\ 1)^T\}$, $\rho_r(A) = 3$, and $\rho_c(A) = 3$.

EXAMPLE 1.2.8. If

$$A = \begin{bmatrix} 1 & 0 & 1 & 0 \\ 0 & 1 & 1 & 0 \\ 1 & 1 & 0 & 0 \\ 1 & 0 & 0 & 0 \end{bmatrix}$$

then $\rho_r(A) = 4$ but $\rho_c(A) = 3$.

DEFINITION 1.2.16. An element x of a lattice L is *join-irreducible* if $x \neq 0$ and $b \vee c = x$ implies $b = x$ or $c = x$. An element of L is *meet-irreducible* if x is not the greatest element and if $b \wedge c = x$ implies $b = x$ or $c = x$. This concept is dual to the concept of join-irreducible.

Note that if $A \in B_{mn}$, then the elements of $B_r(A)$ ($B_c(A)$) are exactly the join-irreducible elements of La R(A) (La C(A)).

We next present three important applications of Boolean matrix multiplication: diagnosis of disease, symbolic logic, and electrical circuit design.

APPLICATION 1.2.1 (diagnosis of disease). R. S. Ledley [227] considers the problem of diagnosing diseases, formalized as follows. Symptoms $S_1, S_2, \ldots, S_n$ and diseases $D_1, D_2, \ldots, D_m$ are considered. A Boolean function $E(S_1, S_2, \ldots, S_n; D_1, D_2, \ldots, D_m)$ describes which symptoms may occur in combination with which diseases. A Boolean function $G(S_1, S_2, \ldots, S_n)$ gives the symptoms presented by a patient. For instance $G = S_1^C S_2$ if the patient has symptom S_2 but not symptom S_1. The diagnosis to be found is a Boolean function $f(D_1, D_2, \ldots, D_m)$ telling which combinations of diseases are consistent with the symptoms.

In this application we write out all combinations of diseases and symptoms and eliminate those that do not agree with E or G. Ledley stated that in general more advanced techniques are needed. These techniques are presumably closely related to the Boolean matrix methods of Applications 1.2.2, 1.2.3.

Another question Ledley considers is that of finding the minimum number of additional tests needed to make an exact diagnosis. In general one may try to find a test that as nearly as possible separates the possibilities into two equal parts. If one is mainly interested in seeing whether or not the patient has a particular disease complex, he gives a procedure which will be illustrated by the following example.

EXAMPLE 1.2.9. By a combination of diseases we mean something like the following: a patient has pneumonia and influenza, or heart disease and kidney disease. Suppose a preliminary examination has narrowed the possibilities for the combination of diseases the patient has down to five. Suppose we make any of four additional tests T_1,

T_2, T_3, T_4. Write out a table as below. The rows are indexed on tests and the columns on combinations of diseases. The (i, j)-entry is 1 if and only if test i is known to come out positive for disease combination j.

Disease Combination

	1	2	3	4	5
T_1	1	0	0	1	0
T_2	0	1	1	0	0
T_3	1	0	0	0	1
T_4	1	1	0	0	1

Suppose we wish to find a procedure which will decide whether the patient has disease complex 1, using as few tests as possible.

If column 1 has all entries 1, go to the next step. Otherwise replace T_i by T_i^C whenevyr the (i, 1)-entry of the table is 0. Here T_i^C denotes the negation of T_i.

Delete column 1. We obtain

	2	3	4	5
T_1	0	0	1	0
T_2^C	0	0	1	1
T_3	0	0	0	1
T_4	1	0	0	1

Next we find those columns with the fewest zeros. This will be column 5. Then find the rows with the fewest ones in those columns. This will be row 1. Thus T_1 will be the first test.

Now delete row 1 and all columns which have zeros in that row.

	4
T_2^C	1
T_3	0
T_4	0

Repeat the process. Column 4 must be selected. Row 3 or row 4 may be selected. Thus the optimum series of tests is Test 1, Test 3 or

Test 1, Test 4. Either will definitely decide whether the patient has disease complex 1.

APPLICATION 1.2.2 (digital computational methods in symbolic logic). R. S. Ledley [226] uses Boolean matrices to find antecedence and consequence solutions of given sets of logical conditions. Suppose $E_t(A_1, A_2, \ldots, A_i; X_1, X_2, \ldots, X_k)$, $t = 1$ to r are a set of logical combinations of the A's and X's which are always true. These are given equations. An antecedence solution is some other set of logical combinations of the A's and X's whose truth implies the truth of the E's. A consequence solution is some set of logical combinations of the A's and X's whose truth is implied by the truth of the E's.

Ledley considers antecedence and consequence solutions of the form $F_u(f_1, f_2, \ldots, f_j; X_1, X_2, \ldots, X_k)$, $u = 1$ to s where F's are given functions and the f's are unknown functions of the A's.

From the functions E, F, form Boolean matrices (e_{ab}) and (f_{ca}), where a runs over the set of possibilities of all values of the X's, b runs over the set of all possibilities of all values of the A's, and c runs over the set of all possibilities of all values of the f's. For given a, b, $e_{ab} = 1$ if and only if all E_t are true for those values of A's and X's and $f_{ca} = 1$ if and only if all F_u are true for those values of the X's and f's.

Then the solution function f can be obtained from the Boolean matrix products

$$(f_{ca})(e_{ab})^C = R^C \quad \text{(antecedence solutions)}$$

$$(f_{ca})^C(e_{ab}) = R^C \quad \text{(consequence solutions)}$$

where C denotes replacing all 0's by 1's and all 1's by 0's.

Ledley also shows that the number of solutions and properties such as logical dependence can be obtained from R. He applies his method to a complicated logical problem in enzyme chemistry and shows how his method might be applied to the genetic code.

EXAMPLE 1.2.10. For $A_1^C X + A_1 A_2 X^C = 0$ search for antecedence solutions of the form $X = f(A_1, A_2)$. We have

(A_1, A_2)

		0 0	0 1	1 0	1 1
E =	X = 0	1	1	1	0
	X = 1	0	0	1	1

X

	0	1
f = 0	1	0
f = 1	0	1

$R^C = FE^C = E^C =$

0	0	0	1
1	1	0	0

Therefore

		0 0	0 1	1 0	1 1
R =	f = 0	1	1	1	0
	f = 1	0	0	1	1

The condition that f is a function of the A's means that we choose one entry from each column of R. For f to be an antecedence solution these entries must be 1.

Therefore we have two choices for f:

	0 0	0 1	1 0	1 1
f = 0	*	*	*	
f = 1				*

	0 0	0 1	1 0	1 1
f = 0	*	*		
f = 1			*	*

In the first case, from the locations of the X's, f = 0 unless $A_1 = A_2 = 1$, so f is A_1A_2. In the second case, f = 0 unless $A_1 = 1$, so f is A_1.

Our solutions are $x = A_1$ and $x = A_1A_2$. Since $A_1^C A_1 + A_1A_2A_1^C = 0 + 0 = 0$ and $A_1^C(A_1A_2) + (A_1A_2)(A_1A_2)^C = 0 + 0 = 0$, these solutions do imply the given equations.

APPLICATION 1.2.3 (digital circuit design). The first use of Boolean algebra in circuit design was made by C. E. Shannon in 1938. Computational difficulties again limited the direct practical application of Boolean algebra. To help offset some of the difficulties new digital computational methods have been developed by R. S. Ledley [227] which have found application to medical science.

Ledley [225] makes use of the techniques of the preceding application. Suppose we wish to design a digital circuit whose output is a given Boolean function of its inputs of the form

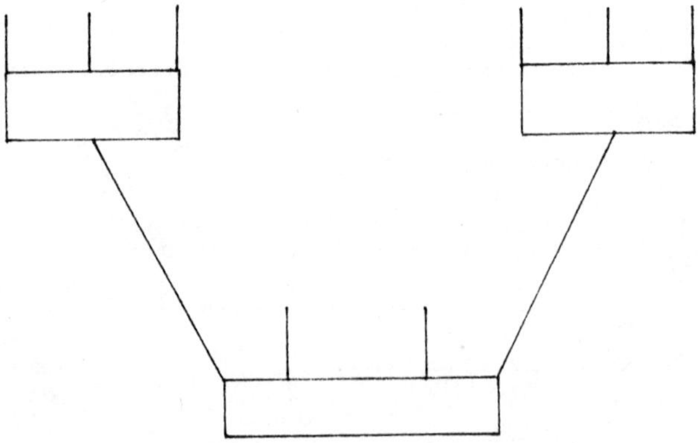

where the A's and X's are inputs, F is a specified circuit, and f_1, f_2 are unknown circuits to be found.

If E represents the desired output, as before, set up the matrices (e_{ab}) and (f_{ca}) where a ranges over the set of all possibilities of values of the X's, b ranges over the set of all possibilities of values of the A's, and c ranges over the set of all possibilities of values of the f's. Let $e_{ab} = 1$ if and only if the function E is 1 for the values of the X's and A's given by a, b and let $f_{ca} = 1$ if and only if the circuit produces a signal for the values of its inputs specified by c, a. Then form the matrices R_1, R_2 defined by $FE^C = R_1^C$, $F^C E = R_2^C$, where the products are Boolean matrix products and C denotes complements.

Then take the intersection R of R_1^C and R_2^C: $r_{ij} = 1$ if and only if both $(R_1)_{ij} = 1$ and $(R_2)_{ij} = 1$. Form a table of possible values of f_1, f_2 from the matrix R, by considering the location of 1 entries in R. This gives the possible solution functions f_1, f_2, which may be translated in terms of circuits.

EXAMPLE 1.2.11. Suppose that it is desired to produce a circuit whose output is given by $E = (A_1^C A_3^C + A_2^C A_3) X_1^C X_2^C + (A_1^C A_3^C + A_1 A_2) X_1 X_2^C + (A_1 A_3^C + A_2 A_3) X_1^C X_2 + (A_1 A_2^C A_3^C + A_1^C A_2 A_3) X_1 X_2$.

Suppose the output of the circuit F is $f_1^C f_2^C X_2^C + f_1 f_2^C X_2 + f_1^C f_2 \cdot (X_1 X_2^C + X_1^C X_2) + f_1 f_2 X_1^C X_2^C$.

The matrix E is

$$\begin{bmatrix} 1 & 0 & 1 & 0 & 1 & 1 & 0 & 0 \\ 1 & 0 & 1 & 1 & 0 & 0 & 0 & 1 \\ 0 & 1 & 0 & 1 & 0 & 0 & 1 & 1 \\ 0 & 1 & 0 & 0 & 0 & 0 & 1 & 0 \end{bmatrix}$$

where the possibilities for the A's have been arranged in the order

$$\begin{bmatrix} 0 & 1 & 0 & 1 & 0 & 1 & 0 & 1 \\ 0 & 0 & 1 & 1 & 0 & 0 & 1 & 1 \\ 0 & 0 & 0 & 0 & 1 & 1 & 1 & 1 \end{bmatrix}$$

and those for the X's in the order

$$\begin{array}{cc} X_1 & X_2 \end{array}$$
$$\begin{bmatrix} 0 & 0 \\ 1 & 0 \\ 0 & 1 \\ 1 & 1 \end{bmatrix}$$

The matrix F is

$$\begin{bmatrix} 1 & 1 & 0 & 0 \\ 0 & 0 & 1 & 1 \\ 0 & 1 & 1 & 0 \\ 1 & 0 & 0 & 0 \end{bmatrix}$$

where the f's are in the order

$$\begin{array}{cc} f_1 & f_2 \end{array}$$
$$\begin{bmatrix} 0 & 0 \\ 1 & 0 \\ 0 & 1 \\ 1 & 1 \end{bmatrix}$$

The matrix R_1, the complement of FE^C, is

$$\begin{bmatrix} 1 & 0 & 1 & 0 & 0 & 0 & 0 & 0 \\ 0 & 1 & 0 & 0 & 0 & 0 & 1 & 0 \\ 0 & 0 & 0 & 1 & 0 & 0 & 0 & 1 \\ 1 & 0 & 1 & 0 & 1 & 1 & 0 & 0 \end{bmatrix}$$

The matrix R_2, the complement of F^CE, is

$$\begin{bmatrix} 1 & 0 & 1 & 0 & 1 & 1 & 0 & 0 \\ 0 & 1 & 0 & 0 & 0 & 0 & 1 & 0 \\ 0 & 0 & 0 & 1 & 0 & 0 & 0 & 1 \\ 0 & 0 & 0 & 0 & 1 & 1 & 0 & 0 \end{bmatrix}$$

The intersection of R_1 and R_2 is (writing out the possibilities on which the rows are indexed)

$$\begin{array}{c} \begin{matrix} f_1 & f_2 \end{matrix} \\ \begin{bmatrix} 0 & 0 \\ 1 & 0 \\ 0 & 1 \\ 1 & 1 \end{bmatrix} \\ \begin{bmatrix} 1 & 0 & 1 & 0 & 0 & 0 & 0 & 0 \\ 0 & 1 & 0 & 0 & 0 & 0 & 1 & 0 \\ 0 & 0 & 0 & 1 & 0 & 0 & 0 & 1 \\ 0 & 0 & 0 & 0 & 1 & 1 & 0 & 0 \end{bmatrix} \end{array}$$

The location of the 1 entries gives the following possibilities for f_1, f_2. For instance in the first column of R a 1 occurs only in the first row, which is the value $f_1 = 0$, $f_2 = 0$. In the second column a 1 occurs only in the second row, which is the case of $f_1 = 1$, $f_2 = 0$. That is

$$\begin{matrix} A_1 \\ A_2 \\ A_3 \end{matrix} \begin{bmatrix} 0 & 1 & 0 & 1 & 0 & 1 & 0 & 1 \\ 0 & 0 & 1 & 1 & 0 & 0 & 1 & 1 \\ 0 & 0 & 0 & 0 & 1 & 1 & 1 & 1 \end{bmatrix}$$

f_1	0	1	0	0	1	1	1	0
f_2	0	0	0	1	1	1	0	1

In this case the functions f_1 and f_2 are unique, and may be written $f_1 = A_1A_2^C + A_1^CA_3$, and $f_2 = A_1A_2 + A_2^CA_3$. Circuits may be constructed to represent these functions.

1.3 Green's Relations

We have mentioned that Boolean matrices under matrix multiplication form a semigroup, which is to say the associative law holds. This fact is of fundamental importance. Over a field most matrices have inverses, and the study of the general linear group is important, but over a Boolean algebra we will later show that only permutation matrices are invertible. Thus the absence of inverses is a fact of great importance.

This fact is also central in semigroup theory. Some way is needed of seeing how close we can come to solving equations like AX = B, XA = B, AXC = B. This is provided by Green's equivalence classes. Green defined five equivalence relations *R*, *L*, *H*, *D*, *J*, in any semigroup, about 1951, and these relations are perhaps the single most important concept in algebraic semigroup theory. The definition of the relation *R*, for instance is that A *R* B if and only if there exist U, V such that AU = B, BV = U. An equivalent definition would be, for all D, AX = D is solvable if and only if BX = D is solvable.

For Boolean matrices these five relationships can be characterized in terms of row and column spaces. The remainder of this section is concerned with advanced combinatorial results: determining the sizes of Green's equivalence classes, and showing that the number of *D*-classes in B_n is asymtotically equal to

$$\frac{2^{n^2}}{(n!)^2}$$

The reader may wish to omit the proofs of these last results the first time he reads the section.

The following definitions of semigroup theory are taken from A. H. Clifford and G. B. Preston [56].

DEFINITION 1.3.1. A *right* (*left*) *ideal* in a semigroup S is a subset X such that XS $\subseteq$ X (SX $\subseteq$ X), and the (*two sided*) *ideal* of S generated by X is SXS $\cup$ XS $\cup$ SX. *Principal ideals*, *principal left* and *right ideals* are defined in a similar way.

DEFINITION 1.3.2. Two elements of a semigroup S are said to be *L-equivalent* if they generate the same principal left ideal of S. *R-equivalence* is defined dually. The *join* of the equivalence relations *L* and *R* is denoted by *D* and their intersection by *H*. Two elements are said to be *J-equivalent* if they generate the same two-sided principal ideal. These five relations are known as *Green's relations*.

Let S be a semigroup. For a, b ε S, we define a L b to mean that a and b generate the same principal left ideal. Other relations are defined in a similar way. Note that L is a right congruence, i.e., a L b implies that ac L bc for all c ε S. Also note that R is a left congruence, i.e., a R b implies that ca R cb for all c ε S.

All H-classes within a D-class have the same number of elements. We can arrange the H-classes within a D-class in a rectangular pattern

$$\begin{array}{cccc} H_{11} & H_{12} & \cdots & H_{1n} \\ H_{21} & H_{22} & \cdots & H_{2n} \\ & \cdots & & \\ H_{m1} & H_{m2} & \cdots & H_{mn} \end{array}$$

in which the rows are the L-classes contained in the D-class and the columns are the R-classes contained in the D-class. An H-class is a group under semigroup product if and only if it contains an idempotent (i.e., $x^2 = x$).

The following results are some simple algebraic characterizations of Green's relations in terms of row and column spaces.

DEFINITION 1.3.3. The *weight* of a vector v, denoted by w(v), is the number of nonzero elements of v. The weight of v is sometimes called the rank of the vector v.

LEMMA 1.3.1 [178]. Two matrices A, B are L (R)-equivalent if and only if they have the same row (column) space.

Proof. Suppose XA = B and YB = A. Then R(B) $\subseteq$ R(A) and R(A) $\subseteq$ R(B) so R(A) = R(B). Suppose R(B) $\subseteq$ R(A). Then by looking at each row of B we can find an X such that XA = B. Likewise we can find a Y such that YB = A. □

LEMMA 1.3.2. Let U be any subspace of V_n and f a homomorphism of commutative semigroups from U into V_n such that f(0) = 0. Then there exists a matrix A such that for all $v \in V_n$, $vA = f(v)$.

Proof. Let $S(i) = \{v \in U: v_i = 1\}$. Then define A_{i*} to be $\inf \{f(v): v \in S(i)\}$. We will show $vA = f(v)$, for all $v \in U$. Suppose $(vA)_j = 1$. Then for some k, $v_k = 1$ and $a_{kj} = 1$. Thus for all $w \in S(k)$, $(f(w))_j = 1$. Since $v \in S(k)$, $(f(v))_j = 1$. This proves that $vA \leq f(v)$.

Suppose that $(vA)_j = 0$. Then for all k such that $v_k = 1$, we have $a_{kj} = 0$. Thus for all k such that $v_k = 1$ we have a vector $x(k) \in S(k)$ such that $f(x(k))_j = 0$. But $(\Sigma\, x(k))_k$ is 1 for each k such that $v_k = 1$. Thus $\Sigma\, x(k) \geq v$. Thus $\Sigma\, f(x(k)) = f(\Sigma\, x(k)) \geq f(v)$. Since $f(x(k))_j = 0$ for each k, $f(v)_j = 0$. This proves $vA \geq f(v)$. □

THEOREM 1.3.3 [244, 413]. Two matrices in B_n belong in the same $\mathcal{D}$-class if and only if their row spaces are isomorphic as lattices.

Proof. The row space of a matrix is the same as its image space on row vectors. And two such spaces are isomorphic as lattices if and only if they are isomorphic as commutative semigroups.

Suppose $A\, \mathcal{D}\, B$. Let C be such that $A\, \mathcal{L}\, C$ and $C\, \mathcal{R}\, B$. Then A, C have identical image spaces on row vectors. Let X, Y be such that $CX = B$ and $BY = C$. Then we have maps $f: V_nC \to V_nB$ and $g: V_nB \to V_nC$ given by multiplying on the right by X and Y. We have fg and gf are the identity. Thus the image spaces of B and C are isomorphic. Thus if $A\, \mathcal{D}\, B$, $R(A) \simeq R(B)$.

Suppose $R(A) \simeq R(B)$. Let h be an isomorphism from V_nA to V_nB. By Lemma 1.3.2 we have matrices X, Y such that $vX = h(v)$ for $v \in V_nA$ and $vY = h^{-1}(v)$ for $v \in V_nB$. Thus XY is the identity on V_nA and YX is the identity on V_nB. Then $A = AXY$ and $V_nAX = V_nB$. Thus $A\, \mathcal{R}\, AX$ and $AX\, \mathcal{L}\, B$. So $A\, \mathcal{D}\, B$. This proves the theorem. □

DEFINITION 1.3.4. Let S be a semigroup, and let $a \in S$. We define: $L_a = \{b \in S: a\ L\ b\}$, $R_a = \{b \in S: a\ R\ b\}$, $H_a = \{b \in S: a\ H\ b\}$, $D_a = \{b \in S: a\ D\ b\}$, and $J_a = \{b \in S: a\ J\ b\}$.

THEOREM 1.3.4 [244, 413]. Let $A \in B_n$. Then the elements of H_A are in one-to-one correspondence with the lattice automorphisms of R(A).

Proof. Let α be an automorphism from R(A) to R(A). Then $A\alpha$ gives a linear map from V_n to V_n sending 0 to 0, i.e., a matrix B. The matrices A, B both have image space R(A), so they are L-equivalent. By Lemma 1.3.2, there exist matrices X, Y so that on R(A) the equations $X = \alpha$ and $Y = \alpha^{-1}$ hold. Then for any vector v it is true that $vAX = vA\alpha = vB$ and $vBY = vB\alpha^{-1} = vA\alpha\alpha^{-1} = vA$. So $AX = B$ and $BY = A$. Thus $A\ H\ B$. This defines a function from the automorphisms of R(A) into the H-class of A. Two different automorphisms will give rise to different maps $A\alpha$ and so to different matrices B. Thus the function is one-to-one. Let B be any matrix in the H-class of A. Then $R(A) = R(B)$ and there exist matrices X and Y such that $AX = B$ and $BY = A$. This implies that X and Y map R(A) to itself and that X gives an automorphism of R(A). This proves the function is onto, and completes the proof of the theorem. □

We now offer some computational results dealing with the size and number of various Green's equivalence classes.

LEMMA 1.3.5 [40]. The number of permutations of n objects with repetitions allowed which may be formed from p objects of which k have been singled out to appear in every one of these permutations is

$$\sum_{i=0}^{k} (-1)^i \binom{k}{i} (p - i)^n$$

Proof. We will prove the lemma by using generating functions for permutations as in J. Riordan [315], although the lemma can be proved directly by analyzing which kinds of permutations are permissible under the rule stated in the lemma.

The generating function for the situation as described in the above will be

$$\left(t + \frac{t^2}{2!} + \ldots\right)\left(1 + t + \frac{t^2}{2!} + \ldots\right)^{p-k} = (e^t - 1)^k (e^t)^{p-k}$$

$$\sum_{i=0}^{k} (-1)^i \binom{k}{i} e^{(k-i)t} e^{(p-k)t} = \sum_{i=0}^{k} (-1)^i \binom{k}{i} e^{(p-i)t}$$

Thus this equation can be written as

$$\sum_{n=0}^{\infty} \sum_{i=0}^{k} (-1)^i \binom{k}{i} (p - i)^n \frac{t^n}{n!}$$

Once we have the generating function, the answer to our problem is the coefficient of $t^n/n!$, which turns out to be

$$\sum_{i=0}^{k} (-1)^i \binom{k}{i} (p - i)^n$$

This completes the proof. □

COROLLARY 1.3.6 [40]. For any integer n,

$$n! = \sum_{i=0}^{n} (-1)^i \binom{n}{i} (n - i)^n.$$

Proof. The right hand side of the equality written above is the number of permutations of n objects (repetitions allowed) from a group of n objects of which n have been singled out to appear at least once in every permutation, but this is precisely the set of all permutations of n objects without repetitions. □

THEOREM 1.3.7 [40]. If $A \in B_n$, $\rho_r(A) = s$, $\rho_c(A) = t$, $|H_A| = h$, and $|R(A)| = |C(A)| = p$, then

(1) $$|L_A| = \sum_{i=0}^{s} (-1)^i \binom{s}{i} (p - i)^n, \quad |R_A| = \sum_{i=0}^{t} (-1)^i \binom{t}{i} (p - i)^n$$

(2) the number of L-classes in D_A is $h^{-1}|R_A|$ and similarly the number of R-classes in D_A is $h^{-1}|L_A|$. The number of H-classes equals the number of L-classes multiplied by the number of R-classes or $h^{-2}|L_A||R_A|$

(3) $|D_A|$ = h(number of L-classes)(number of R-classes) = $h^{-1}|L_A||R_A|$

Proof. Follows directly from structure of Green's equivalence class and Lemma 1.3.5. □

The following corollary is an immediate consequence of Theorem 1.3.7.

COROLLARY 1.3.8. Let $A \varepsilon B_n$ and $\rho_r(A) = \rho_c(A)$. Then $|L_A| = |R_A|$ and the number of L-classes in D_A equals the number of R-classes in D_A.

It is clear from Theorem 1.3.7 that in order to calculate the various numerical quantities associated with a D-class, it is very important to know the size of the H-classes in it.

THEOREM 1.3.9 [244]. For a matrix A let A_0 be the matrix obtained from A by replacing all dependent rows and columns of A, and all but one copy of each independent row and column of A, by zero. Then $A \, D \, A_0$ so $H_A = |H_{A_0}|$. And $H_{A_0} = \{PA_0 : PA_0 = A_0Q$ for some permutation matrices P, Q$\}$.

Proof. There is a natural isomorphism between the row space of A and that of A_0, so A and A_0 are D-equivalent. Thus $|H_A| = |H_{A_0}|$. The indicated set of matrices are both R and L-equivalent to A_0 so they lie in H_{A_0}.

Conversely let $X \varepsilon H_{A_0}$. Then X has the same row space and the same column space as A_0. Suppose $X_{i*} \neq 0$ but $(A_0)_{i*} = 0$. Then some column of X has its i-entry nonzero. But no column of A has i-entry nonzero. This is a contradiction.

Thus X can have no more nonzero rows than A_0. But X must contain all the nonzero rows of A_0, since the nonzero rows are distinct basis vectors. Thus the rows of X are permutations of the rows of A_0. Thus $X = PA_0$ for some permutation matrix P. Likewise $X = A_0Q$ for some permutation matrix Q. This proves the theorem. □

COROLLARY 1.3.10. If A has row or column rank r then $|H_A|$ divides r!.

EXAMPLE 1.3.1. If

$$A = \begin{bmatrix} 1 & 0 & 0 & 0 \\ 0 & 1 & 0 & 0 \\ 1 & 0 & 1 & 0 \\ 1 & 0 & 0 & 1 \end{bmatrix}$$

then clearly $\rho_r(A) = \rho_c(A) = 4$, and $|H_A| = 2$.

Let B denote a $4 \quad 2^4$ matrix in which columns represent all possible column vectors for C(A). Then

$$AB = \begin{bmatrix} 0 & 1 & 0 & 0 & 0 & 1 & 1 & 1 & 0 & 0 & 0 & 1 & 1 & 1 & 0 & 1 \\ 0 & 0 & 1 & 0 & 0 & 1 & 0 & 0 & 1 & 1 & 0 & 1 & 1 & 0 & 1 & 1 \\ 0 & 1 & 0 & 1 & 0 & 1 & 1 & 1 & 1 & 0 & 1 & 1 & 1 & 1 & 1 & 1 \\ 0 & 1 & 0 & 0 & 1 & 1 & 1 & 1 & 0 & 1 & 1 & 1 & 1 & 1 & 1 & 1 \end{bmatrix}$$

and so AB contains exactly 10 distinct column vectors. Thus p = 10. Similarly, B^TA contains exactly 10 distinct row vectors and thus p = 10. Thus

$$|D_A| = \frac{\left(10^n - 4(9)^n + 6(8)^n - 4(7)^n + 6^n\right)^2}{2}$$

THEOREM 1.3.11 [244]. Let A, B $\in B_n$. (1) If A *L* B, then $|C(A)| = |C(B)|$. (2) If A *R* B, then $|R(A)| = |R(B)|$. (3) If A *D* B, then $|R(A)| = |R(B)| = |C(B)| = |C(A)|$.

Proof. Obvious from Theorem 1.2.3 and Lemma 1.3.1. □

We give an asymptotic number which occurs in several forms: $\mathcal{D}$-classes of non-square matrices, families of subsets closed under union, isomorphism types of finite lattices. These results are due to K. H. Kim and F. W. Roush [190].

DEFINITION 1.3.5. A family $\underline{m}$ of subsets of $\underline{n}$ is an *additive space* if $\emptyset \in \underline{m}$ and $\underline{s} \cup \underline{t} \in \underline{m}$ where $\underline{s}, \underline{t} \in \underline{m}$. Two such families are isomorphic if and only if they are isomorphic as semigroups under addition.

DEFINITION 1.3.6. A lattice is of *type* (n, m) if it occurs as some subspace of V_n which is generated (except for 0) under union by m elements. (This is equivalent to saying the lattice has at most m generators other than 0 under union and the lattice has n generators other than the highest element under intersection).

LEMMA 1.3.12. Let n, m tend to infinity in such a way that

$$\frac{\log n}{m} \to 0, \qquad \frac{\log m}{n} \to 0$$

Then the proportion of $m \times n$ matrices which have both row rank m and column rank n tends to 1.

Proof. Let r(i, j) denote the number of matrices with $A_{i*} \geq A_{j*}$ and c(i, j) the number with $A_{*i} \geq A_{*j}$. Then for fixed $i \neq j$ we have

$$\frac{r(i,\ j)}{k} = \left(\frac{3}{4}\right)^m, \qquad \frac{c(i,\ j)}{k} = \left(\frac{3}{4}\right)^m$$

where $k = 2^{nm}$. Thus the number of matrices having no row greater than or equal to any other, and no column greater than or equal to any other is at least

$$\left(1 - (n^2 - n)\left(\frac{3}{4}\right)^m - (m^2 - m)\left(\frac{3}{4}\right)^n\right)2^{nm}$$

All these matrices have row rank m and column rank n. Under the given hypotheses this number divided by 2^{mn} will tend to 1. This completes the proof. □

From this lemma we derive an asymptotic bound on the number of $\mathcal{D}$-classes. Namely for two matrices of row rank n, if they are $\mathcal{L}$-equivalent they have the same row space and the same row basis, but the row basis consists of all rows. Thus the rows of one matrix are a permutation of the other matrix. Thus A $\mathcal{L}$ B if and only if A = PB for a permutation matrix B. Likewise A $\mathcal{R}$ B if and only if A = BQ. It follows that A $\mathcal{D}$ B if and only if A = PBQ. Thus these $\mathcal{D}$-classes have at most n!m! members. Thus the number of $\mathcal{D}$-classes of these row and column rank n element is at least

$$\frac{2^{nm} - o(2^{nm})}{n!m!} = \frac{2^{nm}}{n!m!}\left(1 - O(1)\right)$$

This gives the lower bound. To prove the upper bound, we will prove that the number of $\mathcal{D}$-classes containing a matrix X such that PXQ = X for nonidentity permutation matrices P, Q is

$$o\left(\frac{2^{nm}}{n!m!}\right)$$

This means most most $\mathcal{D}$-classes have at least n!m! members. Thus the number of $\mathcal{D}$-classes cannot exceed

$$\frac{2^{nm}}{n!m!}\left(1 - o(1)\right)$$

Finally we will derive some conditions relating to solutions of PXQ = X. If X were a projective plane, this would mean X has a non-identity collineation. This is an unsolved problem [331].

LEMMA 1.3.13. If P or Q have no more than k cycles the number of solutions X of PXQ = X is no more than 2^{kn} or 2^{km}, respectively.

Proof. Let P have no more than k cycles. Choose one row from each cycle, and specify it. This can be done in 2^{kn} ways, and these rows determine the rest. Similarly for Q. □

LEMMA 1.3.14. If a permutation P has at least k cycles, it will fix at least m - 2(m - k) numbers from $\underline{m}$.

Proof. If i numbers are fixed, $i + 2(k - i) \le m$. □

LEMMA 1.3.15. Let a permutation group G act on a set S of letters. If for any element g of G, g fixes at least $|S| - a$ letters, with $a > 0$, then there is a set of $|S| - 2a + 1$ letters fixed by every element of G.

Proof. The action of G on S gives a linear representation R of G by permutation matrices. Let $o_1, o_2, \ldots, o_k, o_{k+1}, \ldots, o_{k+t}$ be the G-orbits contained in S, where $o_1, o_2, \ldots, o_k$ contain only one element each, and the rest contain more than one element. Corresponding to this orbit decomposition we have a direct sum decomposition $R = R_1 \oplus R_2 \oplus \ldots \oplus R_k \oplus R_{k+1} \oplus \ldots \oplus R_{k+t}$. A theorem in group representation theory [133, p. 280] states that

$$\sum_{g \in G} \operatorname{Tr}(g) = (k + t)|G|$$

But for any $g \in G$, $\operatorname{Tr}(g) \ge |S| - a$ and, assuming $a > 0$, $\operatorname{Tr}(\theta) > |S| - a$, where $\operatorname{Tr}(g)$ denotes the trace of g and θ denotes the identity element of G. Thus $|S| - a < k + t$. Yet $|S| \ge k + 2t$. Thus

$$|S| - a < k + \frac{|S| - k}{2}$$

which yields the desired inequality on k. □

THEOREM 1.3.6 [190]. Let n, m tend to infinity in such a way that n/m tends to a nonzero constant. Then the number of $\mathcal{D}$-classes of $m \times n$ Boolean matrices is asymptotically equal to

$$\frac{2^{nm}}{n!m!}$$

Proof. By Lemma 1.2.12 and the consideration after its proof we need only prove this formula gives an asymptotic upper bound.

Let $k = \sup\{\lim n/m, \lim m/n\}$. Case 1. $\mathcal{D}$-classes containing some X such that $PXQ = X$ for some P, Q such that P has no more than $m - (4k + 1)\log m$ cycles. (All logarithms are base 2). For fixed

P, Q with P satisfying the hypothesis of this case, there are at most

$$2^{(m-(4k+1)\log m)n}$$

matrices X such that PXQ = X, by Lemma 1.3.13. The number of possibilities for P, Q cannot exceed n!m!. Thus the number of possibilities for X in the present case is at most

$$2^{(m-(4k+1)\log m)n}n!m!$$

Thus also the number of $\mathcal{D}$-classes containing at least one such X is at most

$$2^{(m-(4k+1)\log m)n}n!m!$$

The ratio of this number to

$$\frac{2^{nm}}{n!m!}$$

will approach zero.

Case 2. $\mathcal{D}$-classes containing some matrix X such that PXQ = X for some P, Q such that Q has no more than $n - (4k + 1)\log n$ cycles. This case is treated like Case 1.

Case 3. $\mathcal{D}$-classes containing a matrix X such that PXQ = X for some P, Q not both identity but such that PXQ = X does not hold for any P, Q with P having no more than $m - (4k + 1)\log m$ cycles or Q having no more than $n - (4k + 1)\log n$ cycles. For such an X, choose a pair P, Q satisfying PXQ = X such that sup {m - number of cycles in P, n - number of cycles in Q} is a maximum. Let s denote this maximum. We have $0 < s < (4k + 1)\left(\sup \{\log m, \log n\}\right)$. For a given X the set {P: PXQ = X for some Q} forms a group [180]. Each element of this group will fix at least m - 2s letters by Lemma 1.3.14. Thus by Lemma 1.3.12 the whole group will fix at least m - 4s letters. There is a similar group of Q's which fixes at least n - 4s letters.

Fix s. We first choose a set of 4s letters which is to contain the set of all non-fixed letters under {P: PXQ = X for some Q}. There are $\binom{m}{4s}$ such choices. There is $\binom{n}{4s}$ choices for a similar set for {Q: PXQ = X for some P}. Given that these sets are chosen, we can choose P in (4s)! ways to act on its set and Q in (4s)! ways to act on its set. Once P, Q are chosen we can choose X in at most

$$2^{nm-s(\min\{n,m\})}$$

ways by Lemma 1.3.13. Thus for given s, there are at most

$$\binom{m}{4s}\binom{m}{4s}(4s)!(4s)!2^{nm-s(\min\{n,m\})}$$

choices of X having the required value of s. However these X's do not all lie in different D-classes. For any permutation matrices E, F; EXF will lie in the D-class and have the same value of s.

How many different matrices EXF are there for a given X ? We have a group action of the product of two symmetric groups on such matrices, sending Y to EYF^{-1}. Here F^{-1} denotes the inverse of F. The isotropy group of X has order at most (4s)!(4s)! by the remarks above about choosing P, Q such that PXQ = X. Thus D-class containing one X also contains at least

$$\frac{n!m!}{(4s)!(4s)!}$$

other matrices with the same s. Thus the number of D-classes containing matrices of this type for given s is at most

$$\frac{m^{4s}n^{4s}2^{nm-s(\min\{n,m\})}(4s)!(4s)!}{n!m!}$$

Allowing any value of s we have at most

$$\max_{1 \le s \le (4k+1)n_1} \frac{m^{4s}n^{4s}2^{nm-sn_2}((4s)!(4s)!)(4k+1)\log n_1}{n!m!}$$

where $n_1 = \max\{n, m\}$ and $n_2 = \min\{n, m\}$. The ratio of this quantity to

$$\frac{2^{nm}}{n!m!}$$

tends to zero.

Case 4. All PXQ are distinct so the D-classes have at least n!m! elements. There are at most

$$\frac{2^{nm}}{n!m!}$$

D-classes of this type. This proves the theorem. □

COROLLARY 1.3.17. Let k be the number of matrices X such that $PXQ = X$ for some P, Q not both the identity. Then if $n, m \to \infty$ in such a way that $\frac{n}{m}$ approaches a nonzero constant,

$$\frac{k}{2^{nm}}$$

approaches 0.

THEOREM 1.3.18 [190]. Under the hypotheses of Lemma 1.3.12, then the numbers of *D*-classes and *D*-classes of B_{mn} are asymptotically equal to

$$\frac{2^{nm}}{n!}, \qquad \frac{2^{nm}}{m!}$$

respectively. The number of *H*-classes is asymptotically equal to 2^{nm}.

Proof. For an upper bound, for instance for *R*-classes we have

$$\binom{2^m}{n} + \binom{2^m}{n-1} + \cdots + \binom{2^m}{1}$$

for column rank k, by choosing a set of k column vectors to be a column basis. This is less than or equal to

$$\binom{2^m}{n} \sum_{i=1}^{\infty} \left(\frac{n}{2^m - 1}\right)^i$$

which gives the theorem. Similar methods apply in the other cases. This proves the theorem. □

Observe that the above results also give the asymptotic number of subspaces of V_n with m generators (not just isomorphism classes).

1.4 Rank and Combinatorial Set Theory

Rank is as important for a Boolean matrix as it is for matrices over a field. Recall that the row (column) rank of a matrix is the number of rows (columns) in a row (column) basis. However the row rank need not equal the column rank for matrices of dimension larger than 3. There is also a third rank concept: Schein rank, of great importance. Matrices of Schein rank less than or equal to k form a two-sided semigroup ideal, for any k. This is not true for row and column rank, though the row and column rank are invariants of the $\mathcal{D}$-class. We investigate in this section the exact relationships between these three concepts of rank. The main conceptual result is Theorem 1.4.1. However specific calculations depend on advanced work of D. Kleitman in combinatorial set theory (the general problem is still unsolved). We also consider other relationships between combinatorial set theory and sets of Boolean vectors. On a first reading, the reader may wish to skip all proofs except that of Theorem 1.4.1.

In this section we shall point out the close relationship that exists between the rank of Boolean matrices and combinatorial set theory. A number of the results in this section can be interpreted in terms of combinatorial set theory. A good introduction to combinatorial set theory from one viewpoint is P. Erdös and J. Spencer [105].

DEFINITION 1.4.1. For vectors v, w the symbol c(v, w) will denote the matrix $(v_i w_j)$. Such matrices are called *cross-vectors*.

It can be observed that for a matrix A, $\rho_r(A) = 1$ if and only if $\rho_c(A) = 1$ if and only if $a_{ij} = (v_i w_j)$ for some vectors v, w [279].

The following definition is due to B. M. Schein [335].

DEFINITION 1.4.2. The *Schein rank* $\rho_s(A)$ of a matrix A is the least number of cross-vectors whose sum is A.

EXAMPLE 1.4.1. If

$$A = \begin{bmatrix} 1 & 1 & 0 & 0 \\ 1 & 1 & 1 & 0 \\ 0 & 1 & 1 & 1 \\ 0 & 0 & 1 & 1 \end{bmatrix}$$

then $\rho_r(A) = \rho_c(A) = 4$, but $\rho_s(A) = 3$.

DEFINITION 1.4.3. The *maximum rank function*, here denoted by $\rho_f(n)$ is the largest possible rank of a subspace of V_n.

Evaluating the maximum rank function is a problem in extremal set theory [100, 130, 170].

THEOREM 1.4.1 [205]. For any matrix, $\rho_s(A) \le \rho_r(A) \le \rho_f(\rho_s(A))$ and $\rho_s(A) \le \rho_c(A) \le \rho_f(\rho_s(A))$.

Proof. Let $B_r(A) = \{v_1, \ldots, v_k\}$. Let M_i be the matrix whose jth row is v_i if and only if v_i is less than or equal to the jth row of A. Then

$$A = \sum_{i=1}^{k} M_i$$

which implies $\rho_s(A) \le k = \rho_r(A)$. On the other hand, let

$$A = \sum_{i=1}^{k} M_i$$

where the M's are nonzero cross-vectors. Let v_i be a nonzero row of M_i. Then R(A) is contained in the space spanned by the v_i. The space spanned by $v_1, \ldots, v_k$ is a quotient space of V_k. Thus R(A) is a homomorphic image of a subspace of V_k. The rank of this subspace must be at least as large as the rank of R(A), since images of a set of generators give a set of generators. So $\rho_r(A) \le \rho_f(\rho_s(A))$. The second inequalities follow by symmetry. This proves the theorem. □

Note that two equivalent definitions of Schein rank for matrices other than 0, are as follows: (i) ρ_s is the least integer such that A is the product of an $m \times \rho_s$ and an $\rho_s \times n$ matrix, and (ii) the least integer such that R(A) is contained in a space spanned by ρ_s vectors.

Also note that for a regular matrix $\rho_r = \rho_c = \rho_s$ will follow from later results. For a matrix whose row space is a modular lattice $\rho_r = \rho_c$ by a result of R. P. Dilworth [91]. However the matrix $A = (a_{ij}) = 0$ if and only if $i \neq j$ of dimension $\binom{2n}{n}$ has row rank $\binom{2n}{n}$, column rank $\binom{2n}{n}$ and $\rho_s \le 2n$ as we will shortly prove. Let the rows and columns of A be considered to be indexed on all subsets of order n of $\underline{2n}$. Let S_k be the collection of all such subsets containing the number k, for k = 1, 2, ... , 2n. Let $B_k = (b_{ij})$ be the matrix such that $b_{ij} = 1$ if and only if $i \in S_k$, $j \notin S_k$. Then A is the sum of the rank 1 matrices B_k.

THEOREM 1.4.2 [206]. Let p_1, p_2, p_3 be positive integers such that $p_1 \le \rho_f(p_2)$, $p_2 \le \rho_f(p_3)$, $p_3 = \min \{p_1, p_2\}$. Then there exists a matrix $A \in B_n$ for any $n \ge \max \{p_1, p_2\}$ such that $\rho_r(A) = p_1$, $\rho_c(A) = p_2$, $\rho_s(A) = p_3$.

Proof. Assume $p_1 \le p_2$ by symmetry. Let $\{v_1, \dots, v_{p_2}\}$ be a set of p_2 independent vectors in V_{p_1}. Let A_0 be the matrix whose ith column is v_i. Then A_0 is an $p_1 \times p_2$ matrix of column rank p_2. If the Schein rank of A_0 is less than p_1, write A_0 as A_1A_2 where A_2 is a matrix with fewer rows which still has column rank p_2. Continuing in this way we obtain a matrix A_3 with p_2 columns of column rank p_2, such that its Schein rank q equals its number of rows, both being less than or equal to p_1. If we add on any $p_2 - q$ rows to this matrix its column rank can be no greater and no less than p_2.

We will prove the following inductive hypotheses: for $q \le t \le p_2$ there is an $t \times p_2$ matrix of column rank p_2 and Schein rank t. Suppose this hypothesis is valid for t = k. Let A_4 be a minimal such $k \times p_2$ matrix. Suppose there exist k vectors b_i such that the space spanned by the b_i properly contains the space spanned by the rows of a_i of A_4. Let B be the matrix whose ith row is b_i. Then B has Schein rank k and since $CB = A_4$ for some C, B has column rank p_2. If there were some set S_1 of b's such that the set $S_2 = \{a_i$: for some j, $a_i \ge b_j\}$ had fewer elements than S_1, S_1 could be replaced by S_2 giving a

Schein rank less than k. Thus by matching theory [21] there is a matching $b_i \to a_{\pi(i)}$ such that $b_i \le a_{\pi(i)}$ for each i. By applying a permutation to the rows of B we obtain a matrix M such that $M \le A_4$ and M has Schein rank k and column rank p_2. By minimality $M = A_4$. This contradicts the assumption of proper containment on the subspace.

Let u be a vector which for each i has a zero in some location where $a_i = 1$. For $k < p_2$ we may choose u to be nonzero. Let A_5 be the matrix obtained by adding u as a row to A_4. Then A_5 will have column rank p_2 and Schein rank at least k. Suppose A_5 has Schein rank k. Then there are k vectors $v_1, \ldots, v_k$ such that the space spanned by $v_1, \ldots, v_k$ contains $a_1, \ldots, a_k$, and u_k. By the remarks above about $b_1, \ldots, b_k$ being impossible, the v's must equal to the a's in some order. But then it will not be possible to express u_k in terms of the v's. Thus A_5 has Schein rank k + 1. This proves the induction hypothesis. By adding 0 rows and 0 columns that proves the theorem. □

EXAMPLE 1.4.1. It can be shown that $\rho_f(n) \le 2\rho_f(n-1)$ and that for 1, 2, 3, 4 the values of $\rho_f(n)$ are 1, 2, 4, 7.

THEOREM 1.4.3. $\rho_f(n) \ge \binom{n-1}{k} + \rho_f(n-1)$ for any integer $k \le n - 1$.

Proof. Let $\{v_1, \ldots, v_r\}$ be a set of $r = \rho_f(n-1)$ independent vectors in V_{n-1}. Let $u_1, \ldots, u_r$ be the vectors in V_n obtained by adding a final 0 component to the components of v_i. Let S be the set of all vectors in V_n with exactly k + 1 ones including a final one component. Then $S \cup \{u_1, \ldots, u_r\}$ is an independent set in V_n with the required number of elements. □

The inequality

$$\rho_f(n) \le \left(1 + \frac{1}{\sqrt{n}}\right)\binom{n}{[\frac{n}{2}]}$$

was proved by D. J. Kleitman [212] under the weaker assumption that the sum of no two vectors is another in the set, where [x] is the greatest integer less than or equal to x. To indicate his method of proof we prove a weaker version of his result.

THEOREM 1.4.4 [212]. Let V be a subset of V_n such that the weights of the vectors in V assume only two values a and b, and no vector in V is a sum of other vectors in V. Then

$$|V| \le \max \left\{ \binom{n}{b} + \frac{b-a}{b}\binom{n}{a},\ \binom{n}{a} \right\}$$

assuming a < b.

Proof. Let S be the set of all pairs of vectors (u, v) such that u, v ε V_n, u $\le$ v, and u has weight a and v has weight b. For each v ε V,

$$\sum_{u<v} u < v$$

so there exists k ε $\underline{n}$ such that $v_k = 1$,

$$W = \Big(\sum_{u<v} u\Big)_k = 0$$

Here W is the kth component of the vector which is the sum of all vectors u in V which are less than v. The vector v is greater than $\binom{b}{a}$ vectors of weight a. However only $\binom{b-1}{a}$ of these can lie in V, since for vectors of weight a less than v in S, the k component must be zero. Thus for fixed v ε V, at most a proportion

$$\frac{\binom{b-1}{a}}{\binom{b}{a}} = \frac{b-a}{b}$$

of the possible pairs (u, v) in S can have u ε V. Let n_a (n_b) be the number of vectors in V of weight a (b). Then the number of pairs whose lower element is in V is $n_a\binom{n-a}{b-a}$. The number of pairs whose upper element is in V but whose lower element is not is at least

$$\left(1 - \frac{b - a}{b}\right) n_b \binom{b}{a}$$

The total number of pairs in S is $\binom{n}{b}\binom{b}{a}$. Thus

$$\binom{n}{b}\binom{b}{a} \geq n_a \binom{n - a}{b - a} + n_b \binom{b}{a} \frac{a}{b}$$

$$1 \geq \frac{n_a}{\binom{n}{a}} + \frac{n_b}{\binom{n}{b}} \frac{a}{b}$$

We also have

$$0 \leq n_a \leq \binom{n}{a}, \; 0 \leq n_b \leq \binom{n}{b}$$

A short linear programming calculation shows this implies the result of the theorem. This completes the proof. □

D. J. Kleitman obtains a lower bound

$$\left(1 + \frac{1}{n}\right) \binom{n}{[\frac{n}{2}]}$$

in essentially the following way, for n odd. Let $w\left([\frac{n}{2}]\right)$ for $k \in \underline{n}$ consist of all weight $[\frac{n}{2}]$ vectors u such that

$$\sum_{i=0}^{n-1} iu_i \equiv k (\text{mod } n)$$

Then no two vectors in $w\left([\frac{n}{2}]\right)$ differ in less than four places. Thus no two vectors in $w\left([\frac{n}{2}]\right)$ will have sum a vector of weight exactly $[\frac{n}{2}] + 1$. Thus $w\left([\frac{n}{2}]\right)$ together with all weight $[\frac{n}{2}] + 1$ vectors forms a possible set V. At least one of the n sets $w\left([\frac{n}{2}]\right)$ must have

$$\frac{1}{n} \binom{n}{[\frac{n}{2}]}$$

elements, since their union has

$$\binom{n}{[\frac{n}{2}]}$$

elements.

For brevity, we let $m = \frac{n}{2}$. For n even we can obtain a similar bound

$$\binom{n}{m} + \frac{1}{n}\binom{n}{m-1} = \binom{n}{m}\left(1 + \frac{1}{n} + \frac{2}{n(n+2)}\right)$$

If we combine this with Theorem 1.4.3 we do obtain an improved bound for $n = 2k + 1$,

$$\binom{2k}{k} + \binom{2k}{k}\left(1 + \frac{1}{2k} - \frac{2}{k(k+2)}\right)$$

$$= \binom{n}{m}\left(1 + \frac{3}{2n} - 0(n^{-2})\right)$$

DEFINITION 1.4.4. An *antichain* in a poset is a subset such that no element in the subset is greater than any other element in the subset.

We observe that determining $\rho_f(n)$ can be related to the much studied problem of finding a maximum antichain in a poset.

DEFINITION 1.4.5. For a subset S of $2^{\underline{n}}$ and a function $h: S \to \underline{n}$ such that $h(X) \in X$ for all X, $p(S, h)$ is the poset whose elements are the elements of S under the order relation $X < Y$ if and only if $X \subset Y$ and $h(Y) \in X$. Let $g(S, h)$ denote the largest size of an antichain in $p(S, h)$.

PROPOSITION 1.4.5. $\rho_f(n) = \max_{S,h} g(S, h)$

Proof. Let T be a family of sets such that no set of T is a union of properly smaller sets of T. Then let S be T and let h be any function such that

$$h(X) \in X \setminus \bigcup_{\substack{Y \in T \\ Y \subset X}} Y$$

for each X. Then $|T| \le g(S, h)$. Thus $\max_{S,h} g(S, h) \ge T$.

Conversely for any S, h let T be a largest-sized antichain in p(S, h). Then no set X of T can be a union of smaller sets of T, since no smaller set contains h(X). □

THEOREM 1.4.6 [206]. In V_n there exists an independent set of vectors of size $\rho_f(n)$ which contains no vector of weight larger than $\frac{n+1}{2}$ and at most one vector of weight one.

Proof. Let S be an independent set of vectors in V_n with $|S| = \rho_f(n)$ such that the largest weight of a vector in S is as smaller as possible. Let this largest weight be b and let a = b - 1. Let T consist of the vectors in S of weights a and b together with all vectors in V_n of weight a which are sums of vectors in S. Then T is an independent set also. Apply Theorem 1.4.4 to T. We have

$$\binom{n}{a} \geq \binom{n}{b} + \frac{b-a}{b}\binom{n}{a}$$

assuming $b > \frac{n+1}{2}$. Therefore in this case $|T| \leq \binom{n}{a}$, and so we can modify S by replacing the vectors in S of weights a, b with all vectors in V_n of weight a which are not sums of vectors of lower weight in S. Since $|T| \leq \binom{n}{a}$ this will not reduce the size of S and will reduce the largest weight of a vector in S. This contradiction shows that the largest weight of a vector in S must be less than or equal to $\frac{n+1}{2}$.

If there is more than one vector in S of weight one, we can reduce the number by replacing for example (1 0 0) and (0 1 0) by (1 0 0) and (1 1 0). This proves the theorem. □

THEOREM 1.4.7 [206]. The first few values of $\rho_f(n)$ are given by

n	1	2	3	4	5	6
$\rho_f(n)$	1	2	4	7	13	24

Proof. For n = 1, 2, 3 this can be easily be seen. We make use of the preceding theorem together with the following inequalities of D. J. Kleitman [212],

$$\sum_{k=j}^{2j} \frac{n_k}{\binom{n}{k}} \frac{j}{k} \le 1, \quad j = 1, 2, \ldots, [\tfrac{n}{2}]$$

where n_k denotes the number of vectors of weight k in the set.

For n = 4 the result is immediate from the preceding theorem.

For n = 5 one of the inequalities gives $3n_2 + 2n_3 \le 30$. Thus 14 could only be realized in the following ways for $n_4 = n_5 = 0$.

n_1	n_2	n_3
1	3	10
1	4	9

The first of these is impossible because any three weight 2 vectors will have at least one union of weight 3. Suppose the second case holds. We have four weight 2 vectors with only one weight 3 union. Some one of the four will not be less than this union. Thus it will be disjoint from the others, and so these 4 vectors are up to a permutation (1 1 0 0 0), (0 0 1 1 0), (0 0 1 0 1), (0 0 0 1 1). However there is no way to add a weight 1 vector without introducing additional weight 3 unions. So $\rho_f(5) = 14$ is impossible.

For n = 6 the inequality above for j = 2 with the preceding theorem allows only the following possibilities for $\rho_f(n) \ge 25$;

n_1	n_2	n_3
1	5	20
0	5	20
1	4	20
1	5	19
1	6	18

The first four of these can be ruled out by the same methods as for n = 5. Suppose we have the case 1, 6, 18. Suppose one of the weight 2 vectors is disjoint from the others. Then the rest give a set of five weight 2 vectors in V_4 with only two weight 3 unions. Since this is all but one of the weight two vectors in V_4 this cannot happen. So every one of the weight two vectors enters a sum to form one of the two weight 3 unions, and thus the six weight 2 vectors are precisely all weight 2 vectors less than one of the weight 3 unions. If the weight 3 unions overlapped we would have an additional weight 3 union. Thus the weight 3 unions are disjoint. Then up to a permutation the weight 2 vectors are (1 1 0 0 0 0), (0 1 1 0 0 0), (1 0 1 0 0 0), (0 0 0 0 1 1), (0 0 0 1 1 0), (0 0 0 1 0 1). There is no way to add a weight 1 vector without giving another weight 3 union. This contradiction proves $\rho_f(6) = 24$. (The vectors just listed together with all weight 3 vectors except (1 1 1 0 0 0) and (0 0 0 1 1 1) give an independent set with 24 elements). This completes the proof. □

Another question in combinatorial set theory which is of interest in connection with vectors is that answered by Sperner's lemma: *the largest sizes of a family of subsets of* $\underline{n}$ *such that no set in the family is contained in any other set in the family*. One such family is given by all subsets with exactly $[\frac{n}{2}]$ elements. Sperner's lemma asserts that there is no larger family. This was proved by E. Sperner in 1928. Since then a number of other proofs have appeared [45, 115, 213, 252, 406]. In the following we shall formulate the Sperner's lemma in terms of Boolean vectors. This formulation is due to K. H. Kim and F. W. Roush [206].

THEOREM 1.4.8 (Sperner). The maximum number of vectors in V_n such that no vector is less than another is

$$\binom{n}{[\frac{n}{2}]}$$

Proof. Let $S_{w(k)}$ denote the set of vectors in V_n of weight k. We define a function g from $S_{w(k)}$ to $S_{w(k-1)}$ as follows: for a vector v in $S_w(k)$ let a_i denote the number of zeros in v preceding the ith one. Let p be the least positive integer congruent to

$$\sum_{i=1}^{k} a_i \pmod{k}$$

Let g(v) be the vector obtained from v by changing the pth 1 in v to a zero. Suppose g(v) = g(v'). Then each of v and v' equals g(v) except in one place where a zero is replaced by a one. Suppose the change occurs in location x in v and location y in v'. Then location y must be the p'th one in v'. If we write $a_i(v)$ and $a_i(v')$ to distinguish the a's for the two vectors we have (assuming by symmetry that y > x), $a_i(v) = a_i(v')$ unless $p \le i \le p'$, $1 + a_i(v) = a_{i-1}(v')$ for $p < i \le p'$, $a_p(v) = x - p$, and $a_{p'}(v') = y - p'$. Thus, adding these equations

$$\sum_{i=1}^{k} a_i(v) + (p' - p) - (x - p) + (y - p') = \sum_{i=1}^{k} a_i(v')$$

$$\sum_{i=1}^{k} a_i(v) + y - x = \sum_{i=1}^{k} a_i(v')$$

$$p - x \equiv p' - y \pmod{k}$$

$$x - p \equiv y - p' \pmod{k}$$

Since the location x is the pth one in v, x - p is the number of zeros in v preceding that location. Thus the number of zeros preceding the pth 1 in one vector must be congruent to the number of zeros preceding the p'th 1 in the other. Yet the difference z is one greater than the number of zeros properly between locations p and p' in each vector. So $z \equiv 0 \pmod{k}$, $1 \le z \le n - k$ since there are n - k zeros in all. For n - k < k this is a contradiction so g is one-to-one.

Now let C be a maximum sized antichain of vectors. If we apply g to the vectors of highest weight in C we obtain a new antichain with the same number of elements, provided that this weight k > n - k. By repeating this we can ensure the highest weight of a vector in the antichain is no larger than $[\frac{n}{2}]$. By applying a function dual to g to the vectors of lowest weight we can ensure that no vector of weight less than $[\frac{n}{2}]$ occurs. Thus the size of C is at most

$$\binom{n}{[\frac{n}{2}]}$$

This proves the theorem. □

We now proceed to generalize a well-known theorem of P. Turan [105] which usually says: *maximize the number of edges in a graph on n vertices without triangles*. This is equivalent to the least number of vectors of weight k such that every vector of weight k + 1 is greater than at least one of the weight k vectors. Let T(n, k) be this least number with satisfy the above condition. Turan's theorem includes and generalizes the case k = 2.

PROPOSITION 1.4.9 [192]. $\dfrac{T(n, k)}{\binom{n}{k}}$ is nondecreasing in n for fixed k.

Proof. Let $M_{w(k)}$ be a minimal set of n + 1 component vectors of weight k such that every weight k + 1 vector is greater than at least one of them. For i = 1, 2, ... , n + 1, let $V_0 = \{v \in M_{w(k)} : v_i = 0\}$ and $V_1 = \{v \in M_{w(k)} : v_i = 1\}$. We have

$$\sum_{i=1}^{n+1} |V_1| = k|M_{w(k)}|$$

since in the left hand side each element of $M_{w(k)}$ is counted k times. Thus

$$\max_i |V_1| \geq \frac{k}{n+1} T(n+1, k)$$

On the other hand $|V_0| \geq T(n, k)$ and $T(n + 1, k) = |V_1| + |V_0|$. Thus

$$\frac{n+1-k}{n+1} T(n+1, k) \geq T(n, k)$$

This is equivalent to the statement of the proposition. □

THEOREM 1.4.10 [192]. Let $g(k) = \lim_{n\to\infty} \frac{T(n, k)}{\binom{n}{k}}$, then asymptotically in k, $g(k) \le \frac{\log k}{k}$.

Proof. We divide the set of components into j classes $C_1, \ldots, C_j$, as nearly equal as possible. For a weight k vector v let c_i be the number of 1 components it has in C_i. Let S be the set of weight k vectors such that either some $c_i = 0$ or $\Sigma\, ic_i \equiv d(\text{mod } j)$. For any d this set will contain at least one vector less than any weight k + 1 vector: if the weight k + 1 vector has no 1 components in some C_i, remove any of its 1 components. If it has at least one 1 component in each C_i, by removing a 1 component from the proper C_i we can realize all values of $\Sigma\, ic_i$, modulo j.

By choosing d we may assume that the number of vectors with $\Sigma\, ic_i \equiv d(\text{mod } j)$ is less than or equal to $\frac{1}{j}\binom{n}{k}$. The number of vectors with some $c_i = 0$ is less than or equal to $j\binom{n-q}{k}$ where $q = [\frac{n}{j}]$.

This gives, asymptotically,

$$g(k) \le \frac{1}{j} + j\left(1 - \frac{1}{j}\right)^k \simeq \frac{1}{j} + je^{-k/j}$$

Choosing $j = [\frac{k}{\log k}]$ gives the theorem. □

D. E. Rutherford [329] and T. S. Blyth [29] have studied eigenvalues and eigenvectors of a Boolean matrix over an arbitrary Boolean in terms of k-function hypergraphs. For various properties of hypergraphs, see C. Berge [20].

1.5 Eigenvectors

In contrast to the preceding sections, where as eigenvalues and eigenvectors play a major role in the theory of matrices over a field, they play a very minor role in Boolean matrix theory. However for the sake of completeness, in this section we say just what the situation is. The material in this section will not be needed in the rest of the book.

C. Greene [127] noticed that the above theorem can be interpreted values and eigenvectors of a Boolean matrix over an arbitrary Boolean algebra. We will first specialize their results to β_0 and then generalize them to arbitrary commutative semirings containing β_0.

DEFINITION 1.5.1. An *eigenvector* of a matrix A over a commutative semiring R_1 is a vector x such that $xA = \lambda x$ for some $\lambda \in R_1$ or $Ax = \lambda x$ for some $\lambda \in R_1$. The element λ is called the associated *eigenvalue*.

In case of β_0, x is a *row eigenvector* of A if and only if xA = 0 or xA = x. We will call these 0-*eigenvectors*, and 1-*eigenvectors*, respectively.

LEMMA 1.5.1. Let $A \in B_n$ and $m \geq n$. Then $(A + A^2)^m = (A + A^2)^m A$.

Proof. Since $(A + A^2)^m = A^m + \ldots + A^{2m}$ and $(A + A^2)^m A = A^{m+1} + \ldots + A^{2m+1}$, thus it will suffice to prove $A^m \leq A^{m+1} + \ldots + A^{2m+1}$ and $A^{2m+1} \leq A^m + \ldots + A^{2m}$. Supposing $a_{ij}^{(m)} = 1$. Then there exists a sequence i(1), i(2), ..., i(m - 1) such that $a_{i,i(1)}$, $a_{i(1),i(2)}$, ... , $a_{i(m-1),j}$ are all 1. Since $m \geq n$ some pair of integers among i, i(1), ... , i(m - 1), j must be equal. Suppose for instance i(r) and i(s) are equal where s > r. Then take the sequence i, i(1), ... , i(r), ... , i(s), i(r + 1), ... , i(s), i(s + 1), ... , i(m - 1), j. Then $a_{ij}^{(m+r-s)} = 1$. The exponent r - s is at least 1 and no more than m (even in the case where i, j are the pair of integers). Thus $m + 1 \leq m + r - s \leq 2m$. Thus the (i, j)-entry of $(A + A^2)^m A$ is equal to 1. This proves $A^m \leq A^{m+1} + A^{m+2} + \ldots + A^{2m+1}$.

Let $v \in V_n$. Then $0 \leq vA^m \leq v(A^m + A^{m+1}) \leq \ldots \leq v(A^m + \ldots + A^{2m+1})$. If all elements of this chain were distinct then V_n would contain a chain of $m + 2 > n + 1$ vectors. That is impossible. Thus some pair of vectors in the chain are equal. Suppose for instance $v(A^m + \ldots + A^{m+s}) = v(A^m + \ldots + A^{m+s+1})$. Then $vA^{m+s+1} \leq v(A^m + \ldots + A^{m+s})$. Then also $vA^{m+s+2} \leq v(A^{m+1} + \ldots + A^{m+s+1})$. By induction, $vA^{2m+1} \leq v(A^m + \ldots + A^{2m})$. Since v was arbitrary, $A^{2m+1} \leq A^m + \ldots + A^{2m}$. This proves the lemma . □

THEOREM 1.5.2 [29]. Let $A \in B_n$. Then

(1) The set of 0-eigenvectors of A is a subspace of V_n, with basis $\{e_i: A_{i*} = 0\}$.

(2) The set of 1-eigenvectors of A is a subspace, with basis $B_r((A + A^2)^m)$, for $m \geq n$.

(3) Let $x \in V_n$. Let $M(x, \lambda) = \{A \in B_n: xA = \lambda x\}$. Let $M(x) = M(x, 0) \setminus M(x, 1)$. Then $M(x, 0)$, $M(x, 1)$ and $M(x)$ are each subsemigroups of B_n.

(4) The largest matrix in $M(x, 0)$ is $Z = (z_{ij})$ such that $z_{ij} = 1$ if and only if $x_i = 0$.

(5) The largest matrix in $M(x, 1)$ is $W = (w_{ij})$ such that $w_{ij} = x_i^c + x_j$.

Proof. Statement (1) can be proved by a computation, as can (3), (4), and (5).

For (2) we first prove that $xA = x$ if and only if $x(A + A^2)^m = x$. If $xA = x$ then $xA^i = x$ for any i. This implies $x(A + A^2)^m = x$. Suppose $x(A + A^2)^m = x$. Then $xA = (x(A + A^2)^m)A = x(A + A^2)^m = x$, by Lemma 1.5.1.

Next, Lemma 1.5.1 implies that $(A + A^2)^m A^i = (A + A^2)^m$ for any i and this implies $(A + A^2)^m$ is idempotent. For any idempotent E we have $xE = x$ if and only if $x \in R(E)$. Thus $\{x: xA = x\}$ is identical with $R((A + A^2)^m)$. Thus a basis is given by $B_r((A + A^2)^m)$. □

We remark that by symmetry under transpose, the above theorem is also valid for column eigenvectors.

DEFINITION 1.5.2. Let R be a commutative semiring and let $h: \beta_0 \to R$ be a monomorphism sending 0 to an element 0 such that $0 + x = x$, $0x = 0$ for all $x \in R$.

DEFINITION 1.5.3. A matrix M over β_0 *diagonalizes* over the commutative semiring R if there exist matrices A, B over R such that

$$\text{AMB is diagonal} \tag{1.5.1}$$

(1.5.2)
$$BA = \begin{bmatrix} h(1) & 0 & \dots & 0 \\ 0 & h(1) & \dots & 0 \\ & \dots\dots\dots & & \\ 0 & 0 & \dots & h(1) \end{bmatrix}$$

Here h(1) denotes the image of $1 \varepsilon \beta_o$ under the homomorphism h.

DEFINITION 1.5.4. A matrix M over β_o has a *complete set* of eigenvectors if for any two Boolean polynomials $p_1(M)$ and $p_2(M)$ there exists an eigenvector v of M over some semiring such that $p_1(M)v \neq p_2(M)v$.

T. S. Blyth [30] showed that nondiagonal matrices over any Boolean algebra cannot be diagonalized over the Boolean algebra.

THEOREM 1.5.3 [206]. If M is a matrix over β_o with off-diagonal ones, then M does not diagonalize over any commutative semiring R.

Proof. Suppose a commutative semiring R, a homomorphism h, and matrices A, B exist satisfying (1.5.1), (1.5.2). We will replace R by a series of rings the last of which is β_o. Let R_1 be the set of mutiples of h(1) in R, and let A_1, B_1 be A, B after every entry has been multiplied by h(1). Then (1.5.1), (1.5.2) still hold. Let H be a maximal ideal in R_1 and let R_2 be R_1/H, and A_2, B_2 be the homomorphic images of A, B. Then (1.5.1), (1.5.2) still hold. Then R_2 has no ideals other than {0}, R_2. Thus R_2 has no zero divisors. Let $H_1 = \{x \varepsilon R_2 : x + y = 0 \text{ for some } y \varepsilon R_2\}$. Then H_1 is an ideal. Suppose $h(1) \varepsilon H_1$. For some y, $h(1) + y = 0$, $(h(1) + h(1)) + y = h(1) + (h(1) + y)$, $h(1) = 0$ which is false.

Thus H_1 is a proper ideal of R_2. Thus $H_1 = \{0\}$. Thus we obtain a homomorphism from R_2 to β_o by sending 0 to 0 and all nonzero elements to 1. This sends A, B to Boolean matrices satisfying (1.5.1), (1.5.2). However (1.5.2) implies A, B are permutation matrices, $A = B^T$. And $M = A^T DA$ for diagonal D, so M is diagonal. □

THEOREM 1.5.4 [206]. If M is a matrix over β_0 then M has a complete set of eigenvalues.

Proof. Let R_0 be the semiring of polynomials in M over β_0 and let R_1 be the additive semigroup of Boolean vectors on which M acts, on the left. Let R be additively $R_1 \oplus R_0$ and have products defined by $p(M)v = vp(M)$ which is the product of the matrix $p(M)$ and the vector v, $vw = 0$ for $v, w \in R_1$, $p_1(M)p_2(M)$ is the usual product in R_0. Then R is an associative and commutative semiring.

Let u_i represent the element $(0, 0, \ldots, 1, \ldots, 0)^T$ of R_1 which has a 1 only in place i. Let e represent the element M of R_0. Embed β_0 in R by sending 0 to (0, 0) and 1 to (I, 0) where I is the identity matrix. Then $(u_1, u_2, \ldots, u_n)^T$ is an eigenvector of M with eigenvalue e.

Moreover $p_1(M)(u_1, \ldots, u_n)^T = p_2(M)(u_1, \ldots, u_n)^T$ if and only if $p_1(M) = p_2(M)$. This completes the proof. □

1.6 Quadratic Equations

Here we consider quadratic equations for Boolean matrices. In particular finding the square root of a Boolean matrix is a well-known unsolved problem. We show that the probability that a random Boolean matrix has a square root tends to zero as n tends to infinity, and give specific examples of square and cube roots. A sufficient but not necessary condition for solution of a Boolean matrix equation is that the same equation can be solved for (0, 1)-matrices of real numbers. This theory has been extensively investigated in combinatorics, for instance in the theory of projective planes. We state some of the elementary results about solutions of quadratic equations for (0, 1)-matrices of real numbers. The material in this section is not needed for the rest of the book.

Let

(1.6.1) $$AA^T = J$$

(1.6.2) $$A^2 = J$$

(1.6.3) $$A^2 = J \ominus I$$

where $J \ominus I$ denotes the matrix with 0's down its main diagonal and all other entries 1. Two types of matrix multiplication will be considered here, over β_0 and $\mathcal{R}$. The following results are well known.

PROPOSITION 1.6.1. For A a matrix with any real entries all solutions of $AA^T = J$ have each row of A equal to the same vector v of norm 1.

Proof. Let the rows be vectors u_i then $u_i u_j = 1 = u_i u_i$ by Hölder's inequality implies the result. □

Note that every Boolean solution, regarded as a matrix over β_0 is also of this form.

PROPOSITION 1.6.2. For ordinary matrix multiplication, $n > 1$ there are no $\{0, 1\}$ solutions of (1.6.3).

Proof. The eigenvalues of A^2 will be $n - 1, -1, \ldots, -1$. Thus the eigenvalues of A will be $\sqrt{n - 1}, \pm i, \ldots, \pm i$. So the real part of Tr A is nonzero. Here Tr A denotes the trace of A. Yet if Tr A were nonzero, Tr A^2 would be nonzero also. □

DEFINITION 1.6.1. A square matrix A is called a *square root* of B if $A^2 = B$.

No general criterion is known for deciding whether a given matrix either over Boolean algebra or field has a square root nor is there any very quick way to find a square root if one exists, though work in this area has been done by N. G. de Oliveira [267] and others [139, 154, 260].

N. de Oliveira [267] presents a method for finding square roots of Boolean matrices.

DEFINITION 1.6.2. A sequence $A_1, A_2, \ldots, A_n$ of B_n is an *admissible set* if and only if (i) no A_i is identically zero, (ii) in each A_i all nonzero rows are equal, and (iii) the ith row of A_k is zero if and only if kth column of A_i is zero.

EXAMPLE 1.6.1.

$$\begin{bmatrix} 1 & 1 & 0 \\ 1 & 1 & 0 \\ 0 & 0 & 0 \end{bmatrix}, \quad \begin{bmatrix} 1 & 0 & 1 \\ 0 & 0 & 0 \\ 1 & 0 & 1 \end{bmatrix}, \quad \begin{bmatrix} 0 & 0 & 0 \\ 0 & 1 & 1 \\ 0 & 1 & 1 \end{bmatrix}$$

THEOREM 1.6.3 [267]. There is a one-to-one correspondence between square root of a Boolean matrix B, which has no zero rows or columns and admissible systems whose sum is B. To obtain a square root C from an admissible system A_i let the jth row of C be any of the nonzero rows of A_j.

Proof. Suppose A_i is an admissible system such that

$$\sum_{i=1}^{n} A_i = B$$

Let A be defined as in the theorem. Then $c_{ik} \neq 0$ if and only if the kth entry in some row of $A_i \neq 0$ if and only if the kth column of $A_i \neq 0$ if and only if the ith row of $A_k \neq 0$. And $c_{kj} \neq 0$ if and only if the jth entry in some row of $A_k \neq 0$ if and only if the jth column of $A_k \neq 0$. Thus $c_{ik}c_{kj} \neq 0$ if and only if the ith row and the jth column of $A_k \neq 0$ if and only if (i, j)-entry of A_k is equal to 1. Thus

$$c_{ij}^{(2)} = \sum_k (A_k)_{ij}, \qquad C^2 = B$$

Suppose C is a square root of B. Let $(A_k)_{ij} = c_{ik}c_{kj}$. Since B has no zero rows or columns, neither will C. So A_i is nonzero for each i. The set of A's will form an admissible system. Thus $\Sigma\, (A_k)_{ij} = \Sigma\, c_{ik}c_{kj} = c_{ij}^{(2)} = b_{ij}$. It follows that the system has sum B.

It can be verified that the two processes of constructing C from the A_i and the A_i from C are inverse to each other and so give one-to-one correspondence. □

Theorem 1.6.3 is valid over the real field. Let (S, f, g) be a triple where S is a set, f is a mapping of a subset X of the Cartesian product $S \times S$ into S and g is a mapping of a subset Y of the Cartesian product $S \times \ldots \times S$ (n times) into S. Assume S contains at least two elements called zero and one, denoted by θ, α such that (θ, α), (θ, θ), (α, θ), $(\alpha, \alpha) \in X$ and $f(\theta, \alpha) = f(\alpha, \theta) = f(\alpha, \alpha) = \theta$, $f(\alpha, \alpha) = \alpha$. If $s_1, \ldots, s_n \in S$ put $s_1 + \ldots + s_n = g(s_1, \ldots, s_n)$ and $s_i s_j = f(s_i, s_j)$ provided $(s_1, \ldots, s_n) \in Y$ and $(s_i, s_j) \in X$. Under appropriate conditions we can define in a natural way the product of two $n \times n$ matrices over S and the sum of n matrices. Thus Theorem 1.6.3 is valid in the structure (S, f, g).

DEFINITION 1.6.3. Let

$$\left[P = \begin{matrix} 0 & 1 & 0 \ldots 0 & 0 \\ 0 & 0 & 1 \ldots 0 & 0 \\ & \ldots\ldots & & \\ 0 & 0 & 0 \ldots 0 & 1 \\ 1 & 0 & 0 \ldots 0 & 0 \end{matrix} \in P_n \right]$$

Then $A \in B_n$ is a *circulant* if $A = c_0 I + c_1 P + c_2 P^2 + \ldots + c_{n-1} P^{n-1}$, where $I = P^0$ and $c_i \in \beta_0$ for every i.

THEOREM 1.6.4 [206]. The Boolean matrix $J \ominus I$ has a square root if and only if its dimension is at least 7, or is 1.

Proof. Let M be a square root of $J \ominus I$ of dimension less than 7. Than all diagonal entries of M are zero, and at least one of m_{ij}, m_{ji} is zero for each i, j. Thus for each i the three sets $\{i\}$, $\{j: m_{ij} = 1\}$, $\{j: m_{ji} = 1\}$ are disjoint. So one of the latter two sets contains at most two elements. By possibly transposing M, assume the third set has at most two elements.

If $\{j: m_{ji} = 1\} = \{a, b\}$ then $m_{ai} = 1$ and $m_{ji} = 0$ for $j \neq a, b$. Moreover $i \neq a$ and $i \neq b$ since all diagonal entries are zero. Since $M^2 = J \ominus I$ we have $\Sigma\, m_{ax} m_{xi} = 1$. But the only nonzero term of this sum is $m_{ab} m_{bi}$. Thus $m_{ab} = 1$. Also $\Sigma\, m_{bx} m_{xi} = 1$. Its only nonzero term is $m_{ba} m_{ab}$. Thus $m_{ab} = m_{ba} = 1$. But this implies $m_{aa}^{(2)} = 1$ which is false. Likewise the case $\{j: m_{ji} = 1\} = \{a\}$ and is equal to $\emptyset$ are impossible, unless the dimension is 1.

For large values of n, there are circulants which are square roots of $J \ominus I$. If S is the set of 1's in the last row of a circulant M, M^2 will be $J \ominus I$ if and only if $\{a + b: a, b \in S\}$ includes numbers in every congruence class modulo n except 0.

If n is odd and at least 9, $S = \{1, 2, \ldots, [\frac{n}{2}] - 2, [\frac{n}{2}], [\frac{n}{2}] + 2\}$ will do. If $n = 7$, $S = 1, 2, 4$ will do.

If n is even and at least 12, let $S = \{1, 2, \ldots, \frac{n}{2} - 3, \frac{n}{2} - 1, \frac{n}{2} + 2\}$. For $n = 8, 10$ we use a different construction

$$\left[\begin{array}{c|c|c|c} 0 & 1 & 0 & 0 \\ \hline 0 & & & 1 \\ \hline 1 & & & 0 \\ \hline 0 & 0 & 1 & 0 \end{array}\right]$$

where the top and bottom diagonal blocks are 1×1, the edge blocks are entirely 1 or entirely 0 as indicated, and the inner four blocks are circulants. For both 8, 10 we can solve to find appropriate circulants.

For $n = 7$, M is unique up to conjugation by a permutation. However this is not so for large n. This completes the proof. □

THEOREM 1.6.5 [200]. The Boolean matrix $J \ominus I$ has a cube root which is a symmetric circulant whenever its dimension is at least 30.

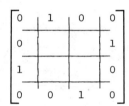

Proof. Let n denote the dimension. If $n \equiv k \pmod 5$, $k = 0, 1, 3$, take the circulant whose last row has ones in locations congruent to ± 1, $\pm(5a + 3)$ for all nonnegative integers a such that $5a + 3 < \frac{n}{2}$.

If $n = 10x + 17$, take the circulant whose row has ones in locations $1, 3, 8, \ldots, 3 + 5(x - 1), 3 + 5x, n - (5 + 5x), n - (3 + 5x), \ldots, n - 1$.

If $n = 10x + 22$, take the circulant whose last row has ones in locations $1, 3, 8, \ldots, 3 + 5(x - 1), 3 + 5x, 5 + 5x, 10 + 5x, n - (10 + 5x), \ldots, n - 1$.

If $n \equiv 4 \pmod 5$, take the circulant whose last row has ones in locations $1, 3, 8, \ldots, 3 + 5(x - 1), 3 + 5x, 5 + 5x, 10 + 5x, n - (10 + 5x), \ldots, n - 1$, where x is the largest integer such that $3 + 5x < \frac{n}{3}$. This proves the theorem. □

In the following we show that the probability that a random Boolean matrix has a square root tends to zero as n goes to infinity.

DEFINITION 1.6.4. The term *random Boolean matrix* refers to a random variable which assumes any matrix in B_n with probability 2^{-n^2}.

DEFINITION 1.6.5. Let X be a random variable which assumes m distinct values $X_1, \ldots, X_m$ and let $p_1, \ldots, p_m$ be the probabilities of these respective values. Then the *entropy* of X is

$$\sum_{i=1}^{m} p_i \log p_i$$

We will use the theorem in entropy that if $Y_1, \ldots, Y_k$ are any random variables, not necessarily independent, the entropy of $(Y_1, \ldots, Y_k)$ is less than or equal to the sum of the entropies of the Y_i. We apply this in the case that $k = n^2$, each Y_i is a particular entry of a random Boolean matrix and $(Y_1, \ldots, Y_k)$ is the Boolean matrix itself (written in an unusual way).

LEMMA 1.6.6. The probability that the (i, j)-entry of A^2 is 1 is $1 - (\frac{5}{8})(\frac{3}{4})^{n-2}$ if $i \neq j$, $1 - (\frac{1}{2})(\frac{3}{4})^{n-1}$ if $i = j$. (Here the Boolean matrices are given the equiprobable distribution).

Proof. If $i \neq j$, we may assume $i, j = 1, 2$. Then partition A as

$$\begin{bmatrix} A_1 & A_2 & A_3 \\ A_4 & A_5 & A_6 \\ A_7 & A_8 & A_9 \end{bmatrix}$$

where A_1 and A_5 are 1×1 matrices and A_9 is an $(n - 2) \times (n - 2)$ matrix. Then divide into two cases according to $A_2 = 0$ or $A_2 = 1$.

For $i = j$, assume $(i, j) = (1, 1)$ and partition A into four blocks. This proves the lemma. □

THEOREM 1.6.7 [200]. The probability that a random $n \times n$ Boolean matrix is a square is $O(n(\frac{3}{4})^n)$.

Proof. By Lemma 1.6.6, the entropy of a particular entry of A^2 is $O(n(\frac{3}{4})^n)$, so the entropy of the random variable A^2 is $O(n^3(\frac{3}{4})^n)$. For any random variable on $n \times n$ matrices, every possible value of that variable has probability at least 2^{-n^2}. This implies that if a random variable on $n \times n$ matrices has m possible values, its entropy is at least $(m - 1)n^2 2^{-n^2}$ (this follows by a consideration of the sum defining entropy). Thus the random variable A^2 assumes at most $O(n(\frac{3}{4})^n 2^{n^2})$ values. This proves the theorem. □

We turn to equation (1.6.1), with ordinary matrix multiplication. These results are well known.

THEOREM 1.6.8. For a solution (1.6.2) to exist over $\{0, 1\} \subset Z$, dimension must be the square of an integer k, and every row and every column sum to k.

Proof. Assume $AJ = JA$. Then this implies all rows and columns have the same sum s and $AJ = sJ$. Then looking at A^2J shows s^2 is equal to the dimension J. This proves the theorem. □

It is frequently convenient to partition A into k^2 $k \times k$ blocks. one solution is given by $a_{xk+y,zk+w} = 1$ if and only if $x + 1 = w$ where $x, z \in \{0, 1, \dots, n - 1\}$ and $y, w \in \underline{n}$.

LEMMA 1.6.9. Let A be a solution to (1.6.2) over $\{0, 1\} \subset Z$. Then some conjugate of A by a permutation matrix has $a_{xk+y,zk+y} = 1$ whenever $z = 0$, $x + 1 = w$ and whenever $y = 1$, $x + 1 = w$.

Proof. Since the eigenvalues of A are $k, 0, \dots, 0$, $\text{Tr } A = k$. Choose a column which contains a diagonal entry, and by conjugation move it to be the first column. So the (1, 1)-entry is 1. Next conjugate by a permutation fixing 1 so that the other k entries in this column are moved to places $2, 3, \dots, k$. Then on multiplying all rows by this column, the first a columns have their sets of 1's disjoint from each other. In turn move the ones of the second column to places $k + 1, k + 2, \dots, 2k$ and so on, fixing all previous locations. Thus the first a columns will have the correct form

$$\begin{bmatrix} 1 & 0 & 0 & \cdots & 0 \\ 1 & 0 & 0 & \cdots & 0 \\ 1 & 0 & 0 & \cdots & 0 \\ 0 & 1 & 0 & \cdots & 0 \\ 0 & 1 & 0 & \cdots & 0 \\ 0 & 1 & 0 & \cdots & 0 \\ 0 & 0 & 1 & \cdots & 0 \\ 0 & 0 & 1 & \cdots & 0 \\ 0 & 0 & 1 & \cdots & 0 \\ & & \cdots\cdots & & \end{bmatrix}$$

Now multiply every row by these columns. Thus each row has 1 entry per block. The remaining conjugations will affect only the positions within blocks, sending each block to itself. By such a conjugation we may locate the entries in row 1 in their proper places. Multiplication by row 1 shows the ones in rows 1, k + 1, 2k + 1, ... , (k - 1)k + 1 are all distinct. So keeping the ones in row 1 fixed move the ones in row k + 1 to their locations, and so on. □

THEOREM 1.6.10. Every solution in the form of Lemma 1.6.9 arises from a ternary product f(a, b, c) on setting (a, b), (c, f(a, b, c)) as the locations of the ones. Such a ternary product will give a solution A if and only if f(x, f(a, b, x), d) is a one-to-one function of x for any a, b, d. A solution is conjugate to the standard solution if and only if all entries of AA^T either equal to 0 or a.

Proof. The first statement means in any row there is exactly one 1 in each block. This follows on multiplying by the first a column. The hypothesis of the last statement implies any rows of A are equal or disjoint. This with the form of Lemma 1.6.9 implies the solution is standard. The second statement is computation. This proves the theorem. □

Many different formulas in fact give the standard solution. For finding nonstandard solutions it seems easier to start with Lemma 1.6.9 and try filling in 0's and 1's than to use ternary rings.

EXAMPLE 1.6.2.

$$\left[\begin{array}{ccc|ccc|ccc}
1&0&0&1&0&0&1&0&0\\
1&0&0&1&0&0&0&0&1\\
1&0&0&0&1&0&1&0&0\\
\hline
0&1&0&0&1&0&0&1&0\\
0&1&0&0&1&0&0&1&0\\
0&1&0&1&0&0&0&1&0\\
\hline
0&0&1&0&0&1&0&0&1\\
0&0&1&0&0&1&1&0&0\\
0&0&1&0&0&1&0&0&1
\end{array}\right]$$

Some interesting work on matrix equations is given in P. Erdös and A. Renyi [103], as well as additional references.

Exercises

1. Prove that $\{\beta_0, +\}$ and $\{\beta_0, \cdot\}$ are semigroups.
2. Prove that $\{\beta_0, +, \cdot\}$ is a semiring.
3. Find a basis of each of the following subspaces of V_3:
   (1) $S_1 = \{(0\ 0\ 0), (1\ 0\ 0), (0\ 1\ 0), (1\ 1\ 0)\}$
   (2) $S_2 = \{(0\ 0\ 0), (1\ 0\ 1), (1\ 1\ 0), (0\ 1\ 1), (1\ 1\ 1)\}$
   (3) $S_3 = \{(0\ 0\ 0), (0\ 0\ 1), (0\ 1\ 1), (1\ 1\ 1)\}$
4. Determine the various subspaces of V_3.
5. Determine the various distributive sublattices of V_3.
6. Determine the various lattices of subspaces of V_3.
7. Determine the (i) row basis, (ii) column space, (iii) row rank of the following matrices:

$$\begin{bmatrix} 1 & 0 \\ 0 & 1 \end{bmatrix}, \quad \begin{bmatrix} 1 & 0 & 0 \\ 1 & 1 & 0 \\ 0 & 0 & 1 \end{bmatrix}, \quad \begin{bmatrix} 1 & 0 & 1 & 0 \\ 0 & 1 & 1 & 0 \\ 1 & 1 & 0 & 0 \\ 1 & 0 & 0 & 0 \end{bmatrix}$$

8. For the matrix

$$\begin{bmatrix} & 1 & 2 & 3 & 4 & 5 & 6 \\ T_1 & 1 & 0 & 1 & 1 & 0 & 0 \\ T_2 & 0 & 1 & 1 & 0 & 1 & 0 \\ T_3 & 1 & 0 & 0 & 0 & 1 & 1 \\ T_4 & 0 & 1 & 1 & 0 & 0 & 1 \\ T_5 & 0 & 1 & 0 & 0 & 1 & 0 \\ T_6 & 1 & 0 & 0 & 1 & 0 & 0 \end{bmatrix}$$

   What is the least number of tests which will definitely tell whether the patient has disease combination 5 ?
9. For AB + AX + BX search for antecedence conditions of the form $X = f(A, B)$.
10. Find 3×3 Boolean matrix A such that $A^3 = 0$ but $A^2 \neq 0$.

11. Prove that if $I \le X$ and $I \le Y$ then $XY \ge X + Y$. Give examples where equality does and does not hold.
12. Draw a diagram for La $R(I_3)$ and La $C(I_3)$.
13. Prove that if a Boolean matrix has row rank n, its row space contains at most 2^n elements. Give an example. Can its row space have $2^n - 1$ elements ?
14. Prove $\{B_n, +, \cdot\}$ is a semiring.
15. Write out all the $\mathcal{D}$-classes of B_3.
16. Write out all the elements of $\mathcal{H}$-class H_{I_3}.
17. Compute $|D_A|$ of

$$A = \begin{bmatrix} 1 & 0 & 1 \\ 0 & 1 & 0 \\ 1 & 0 & 1 \end{bmatrix}$$

18. Prove that a lattice is isomorphic to the row space of a matrix in B_n if and only if it has at most n nonzero generators and its dual has at most n nonzero generators.
19. Prove that if $|R(AB)| = |R(B)|$ then $AB \,\mathcal{L}\, B$. Prove that if $|R(AB)| = |R(A)|$ then $AB \,\mathcal{R}\, A$. Prove that if $|R(A^2)| = |R(A)|$ then $A \,\mathcal{H}\, A^n$ for any n.
20. Do Theorems 1.3.1, 1.3.3 hold for infinite Boolean matrices ? For infinite Boolean matrices give an example to show $\mathcal{J}$-class and $\mathcal{D}$-classes need not coincide.
21. Let

$$A = \begin{bmatrix} 1 & 0 & 0 \\ 1 & 1 & 0 \\ 1 & 0 & 1 \end{bmatrix}$$

Use Theorem 1.3.4 to find $|H_A|$.
22. Prove Lemma 1.3.5 by the inclusion-exclusion formula.
23. Compute the value of $\rho_f(n)$ for n = 8.
24. Prove that the Schein rank of I_n^C, the complement of the identity matrix is k if

$$n = \binom{k}{[\frac{k}{2}]}$$

25. Determine the Schein rank of each of the following matrices:

$$\begin{bmatrix} 1 & 0 \\ 1 & 0 \end{bmatrix}, \quad \begin{bmatrix} 1 & 0 & 0 \\ 1 & 1 & 0 \\ 1 & 1 & 1 \end{bmatrix}, \quad \begin{bmatrix} 1 & 0 & 0 & 1 \\ 0 & 0 & 1 & 0 \\ 0 & 1 & 0 & 1 \\ 1 & 0 & 1 & 0 \end{bmatrix}$$

26. Find the eigenvalues of

$$\begin{bmatrix} 1 & 0 & 0 \\ 1 & 1 & 0 \\ 1 & 1 & 1 \end{bmatrix}$$

27. Compute the square root of

$$\begin{bmatrix} 1 & 0 & 1 \\ 0 & 0 & 0 \\ 1 & 0 & 1 \end{bmatrix}$$

28. Compute the square root of J_4.

29. Show that the equation BA = AB + I implies AB must have at least one main diagonal 1 entry. Characterize all solutions where AB has only one main diagonal 1 entry.

30. Prove that (i) $T(2m, 2) = 2\binom{m}{2}$, and (ii) $T(2m + 1, 2) = m^2$.

31. Find the number of $n \times n$ Boolean matrices having no zero columns or zero rows.

32. Which of these Boolean matrices have square roots ?

$$\begin{bmatrix} 1 & 0 & 0 \\ 1 & 1 & 0 \\ 0 & 1 & 1 \end{bmatrix}, \begin{bmatrix} 1 & 1 & 0 \\ 1 & 1 & 1 \\ 1 & 1 & 1 \end{bmatrix}, \begin{bmatrix} 1 & 0 & 0 \\ 0 & 1 & 0 \\ 0 & 1 & 0 \end{bmatrix}, \begin{bmatrix} 0 & 0 & 1 \\ 0 & 0 & 1 \\ 1 & 1 & 1 \end{bmatrix}, \begin{bmatrix} 0 & 1 & 1 \\ 0 & 0 & 1 \\ 1 & 1 & 1 \end{bmatrix}$$

CHAPTER TWO

ALGEBRAIC PROPERTIES

In this chapter we shall discuss the essential algebraic properties of Boolean matrices.

2.1 Regular Elements and Idempotents

Regular elements are of interest because of their close relationship with inverses which we shall investigate in the next chapter. The notion of regularity was first introduced by J. von Neumann [266] for rings. Since this section is related to the Green's equivalence classes, it is suggested that the reader return to Section 1.3 and review the definitions and Lemmas 1.3.1, 1.3.2 and Theorems 1.3.3, 1.3.4. Most of the results in this section can be found in G. Markowsky [244].

DEFINITION 2.1.1. Let a, b be elements of a semigroup S. We say that a is a *regular element* of S (or simply *regular*) if and only if there exists $x \in S$ such that $a = axa$. Let Reg (X) denote the set of all regular elements contained in a set X.

DEFINITION 2.1.2. Let a be element of a semigroup S. We say that a is *idempotent* if $a^2 = a$. We note that if a is idempotent, then it is regular. Let Idem (X) denote the set of all idempotents contained in a set X.

DEFINITION 2.1.3. Two elements a and b of a semigroup S are said to be *inverses* of each other if aba = a and bab = b.

In any semigroup these three concepts are closely related. An element is regular if and only if it has an inverse if and only if it lies in the same D-class as some idempotent.

Idempotent Boolean matrices are important in the majority of applications of Boolean matrix theory. They occur in a wide variety of forms, yet many general theorems describe their structure. Theorem 2.1.20 of B. M. Schein describes them in terms of quasi-orders. Theorem 2.1.25 of K. A. Zaretski states that a Boolean matrix is regular if and only if its row space froms a distributive lattice. Kim and Roush have related regularity to identification vectors (see Lemma 2.1.24). These can be found by examining the matrix. Using Theorem 2.1.26 the number of idempotents in a D-class can be counted.

The earlier part of this section is an exposition of the theory of idempotents in general semigroups and the method of simplifying an idempotent by deleting dependent rows and columns. By deleting dependent rows and columns from any idempotent we can obtain a D-equivalent idempotent which is a *reduced idempotent*, that is, every nonzero row vector is a row basis vector. A D-equivalent idempotent is the direct sum of a zero matrix and the matrix of a partial order relation. Many other results on idempotent matrices can be found in other portions of the book.

Despite its difficulty, we recommend that the reader go through this section in detail, except possibly for the later combinatorial results. It is central to much of the rest of the book.

We will first discuss some of the general properties of regular elements and idempotents. The following proposition shows that in any semigroup every D-class consists entirely of regular elements or entirely of irregular elements, every element lies in the L-class of an idempotent and the R-class of an idempotent, that to a pair of inverses is associated a pair of idempotents, and that an H-class contains at most one idempotent.

PROPOSITION 2.1.1. Let a, b be elements of a semigroup S.

(1) If $a \in \text{Reg}(S)$, then $b \in \text{Reg}(S)$ if $b \in D_a$.

(2) If $a \in \text{Reg}(S)$, then every L-class and every R-class contained in D_a contains an idempotent.

(3) If $a \in \text{Idem}(S)$, then $xa = x$ for all $x \in L_a$, $ax = x$ for all $x \in R_a$ and $xa = x = ax$ for all $x \in H_a$.

(4) If a and b are inverses of each other, then if we let $e = ab$ and $f = ba$ we have that $e, f \in \text{Idem}(S)$ such that $ea = af = a$ and $be = fb = b$. Thus $e \in R_a \cap L_b$ and $f \in R_b \cap L_a$ and we have that $a, b, e, f \in D_a$.

(5) If $e, f \in \text{Idem}(S)$ and $e\ H\ f$, then $e = f$.

Proof. (1) and (2) Let $x \in S$ be such that $axa = a$. Let $e = ax$ and $f = xa$. It follows that $e^2 = e$ and $f^2 = f$. By Lemma 1.3.1, $e \in R_a$ and $f \in L_a$, and so the L-class and the R-class of a regular element contain an idempotent. Let $b \in D_a$ and so there exist $x_1, y_1, x_2, y_2 \in S$ such that $b = x_1 a x_2$, $a = a x_2 y_2$, $a = x_1 y_1 a$. Consider $b(y_2 x y_1)b = x_1(a x_2 y_2)x(y_1 x_1)a x_2 = x_1(axa)x_2 = x_1 a x_2 = b$, so that b is regular.

(3) If $x \in L_a$, then there exists $z \in S$ such that $za = x$, this implies $xa = (za)a = za = x$. The result for R_a follows similarly and so we get the result for H_a.

(4) If a and b are inverses of each other, then $a\ D\ b$ by Theorem 1.3.3. The rest follows from Lemma 1.3.1.

(5) Follows directly from (3) since we would have $a = ab = b$. This proves the proposition. □

The following interesting lemma is known as Green's lemma and can be found in A. H. Clifford and G. B. Preston [56], although in slightly different form.

LEMMA 2.1.2 [127]. Let a, b, x_1, y_1, x_2, y_2 be elements of a semigroup S.

(1) If $x_1 a = b$ and $y_1 b = a$, i.e., $a\ L\ b$, then there exists an L-class preserving bijection between R_a and R_b, i.e., all R-classes

contained in a D-class have the same cardinality.

(2) If $ax_1 = b$ and $by_1 = a$, i.e., a R b, then there exists an R-class preserving bijection between L_a and L_b, i.e., all L-classes contained in a D-class have the same cardinality.

(3) If $x_1ax_2 = b$, $y_1x_1a = a$, $ax_2y_2 = a$, i.e., a D b, then there exists a bijection between H_a and H_b, i.e., all H-classes contained in a given D-class have the same cardinality.

Proof. (1) Let $f: R_a \to S$ be given by $f(w) = x_1w$ for all $w \in R_a$. Since R is a left congruence $f(w) \in R_b$ (since $b = x_1a$). Thus we can actually consider f to be a map into R_b. Similarly define $g: R_b \to R_a$ by $g(w) = y_1w$. We claim $g \circ f$ is an identity on R_a. Let $w \in R_a$, then there exists a $z \in S$ such that $az = w$, and so $g(f(w)) = g(f(az)) = (y_1x_1a)z = az = w$. Similarly, $f \circ g$ is an identity on R_b. We have therefore shown the existence of a bijection between R_a and R_b and now we need only establish that f is an L-class preserving map. Let $w, z \in R_a$ be such that $w \in L_z$, then there must exist t_1, t_2 such that $t_1w = z$ and $t_2z = w$. But $f(w) = x_1w$ and $f(z) = x_1z$. Let $t_1' = x_1t_1y_1$ and let $t_2' = x_1t_2y_1$, then we have $t_1'f(w) = f(z)$ and $t_2'f(z) = f(w)$ since $g \circ f$ is the identity. In essence this means that $f(H_a) = H_b$ and that in particular $|H_a| = |H_b|$.

(2) The proof is similar to the proof of (1).

(3) From the proofs of (1) and (2) above if $c \in S$ is such that a L c or a R c, then there exists a bijection between H_a and H_b. So (3) follows from this. This proves the lemma. □

The next interesting theorem is called Green's theorem and can be found in A. H. Clifford and G. B. Preston [56].

THEOREM 2.1.3 [127]. Let a, b be elements of a semigroup S such that $a, b, ab \in H_a$. Then H_a is a group. In particular, any H-class which contains an idempotent is a group. Thus an H-class is a group if and only if it contains an idempotent.

Proof. From Lemma 2.1.2 and the fact that $b \in L_{ab}$ we get that the map $f: H_a \to H_a$ given by $f(x) = ax$ is one-to-one and onto and similarly that the map $g: H_a \to H_a$, where $g(x) = xb$, is also bijective. Thus $aH_a = H_a = H_a b$, which means for any $c \in H_a$, $ac \in H_a$ and $cb \in H_a$, and so by repeating the above arguments we see that for all $c \in H_a$ we have that $cH_a = H_a = H_a c$, i.e., for every $c, d \in H_a$ there exist $x, y \in H_a$ such that $cx = yc = d$. We claim that H_a is a group. Let c, d, x, y be such that $cx = d$ and $dy = c$. Let $e = xy$. Clearly $ce = c$. For any $w \in H_a$ there exists a $z \in H_a$ such that $w = zc$, and so $we = zce = zc = w$, and e is a right-identity for H_a. Every element of H_a has a right-inverse and so H_a is a group since it is fairly easy to show that a set which has a right-identity and a right-inverse for every element is actually a group. This proves the theorem. □

We will show that the row rank of any regular matrix is equal to its column rank. We may assume throughout the following discussion that $A \neq 0$, or equivalently that 0 is not the only element of $R(A)$ and of $C(A)$.

The next six lemmas due to G. Markowsky [244, 245] describe idempotents. If $a_{ii} = 0$ then row i is dependent on the other rows, and column i is dependent on the other columns. Deleting this row and column will not change row or column rank (in fact will not change D-class), and the new matrix will be idempotent. If $a_{ii} = 1$ for all i but row i is dependent on the other rows, then also column i is dependent on the other columns, and in fact there exists j such that $A_{i*} = A_{j*}$, $A_{*j} = A_{*i}$. Again we can delete row and column i without changing idempotency or row and column rank. As a consequence we are able to prove that for all regular matrices row, column, and Schein rank are equal.

REMARK 2.1.1. If $A \in B_n$, then $A \in \text{Idem}(B_n)$ if and only if $a_{ij} = 1$ if and only if there exists a $k \in \underline{n}$ such that $a_{ik} = a_{kj} = 1$. This statement follows from the fact that since A is an idempotent we must have that $a_{ij} = \Sigma\, a_{it}a_{tj}$ for all $i, j \in \underline{n}$. Another way of writing this is to observe that $A \in \text{Idem}(B_n)$ if and only if $A_{i*} = \sum_{\mu} A_{j*}$, where $\mu = \{k: e_k \le A_{i*}\}$ for all i and $A_{*i} = \sum_{\nu} A_{*j}$, where $\nu = \{k: e^k \le A_{*i}\}$ for all i.

LEMMA 2.1.4. If $A \in \text{Idem}(B_n)$ and $A_{*i} \in B_r(A)$ then $A_{i*} = A_{t*}$ such that $a_{tt} = 1$. Therefore if $a_{ii} = 0$, then A_{i*} is dependent on the other rows. The same holds true for column vectors.

Proof. We have $A_{i*} = \sum_{\theta} A_{j*}$ where $\theta = \{j: a_{ij} = 1\}$. Since A_{i*} a row basis vector some element A_{t*} of the sum must equal A_{i*}. By definition of θ, $a_{it} = 1$. Thus $a_{tt} = 1$ since $A_{i*} = A_{t*}$. □

The next two lemmas are just a restatement of facts that were dealt with previously.

LEMMA 2.1.5. Let $A \in \text{Idem}(B_n)$ and $a_{tt} = 0$.

(1) If $a_{st} = 1$, then there exists $h \in \underline{n} \setminus \{t\}$ such that $a_{sh} = a_{ht} = 1$.

(2) If $a_{st} = a_{sh} = a_{ht} = 1$, then $A_{*t} + A_{*h} = A_{*t}$, i.e., $A_{*h} \le A_{*t}$. Of course a similar result holds for rows of A.

LEMMA 2.1.6. Let $A \in \text{Idem}(B_n)$ and $a_{tt} = 0$. If $B \in B_{n-1}$ is a matrix formed from A by removing all the elements of A_{t*} and A_{*t}, then (i) $\rho_r(A) = \rho_r(B)$ and $\rho_c(A) = \rho_c(B)$, (ii) $B \in \text{Idem}(B_n)$.

Proof. (i) By Lemma 2.1.4, A_{*t} is dependent on the other columns of A and A_{t*} is dependent on the other rows of A. Thus, it is not hard to deduce that $\rho_r(A) = \rho_r(B)$ and $\rho_c(A) = \rho_c(B)$.

(ii) Let $f: \underline{n-1} \to \underline{n} \setminus \{t\}$ be such that $f(i) = i$ if $i < t$ and $f(i) = i + 1$ if $i \ge t$. Then $b_{ij} = a_{f(i),f(j)}$ for all $i, j \in \underline{n-1}$. To show that $B \in \text{Idem}(B_n)$ we must show that

$$b_{ij} = \sum_{\underline{n-1}} b_{ip} b_{pj}$$

for all i, j ε $\underline{n-1}$. Let

$$c_{ij} = \sum_{\underline{n-1}} b_{ip}b_{pj} = \sum_{p<t} a_{f(i),p}a_{p,f(j)}$$

$$+ \sum_{p>t} a_{f(i),(p+1)}a_{(p+1),f(j)}$$

Thus we see that $a_{f(i),f(j)} = c_{ij} + a_{f(i),t}a_{t,f(j)}$ since A ε Idem (B_n). Thus to show that B ε Idem (B_n), we must show that $c_{ij} \geq a_{f(i),f(j)}$ or equivalently that $c_{ij} = 0$, implies that $a_{f(i),f(j)} = 0$, i.e., $a_{f(i),t}a_{t,f(j)} = 0$. What we will show is that if $a_{f(i),t}a_{t,f(j)} = 1$, then $c_{ij} = 1$. To simplify notation, let m = f(i) and p = f(j). Thus we have that $a_{mt}a_{tp} = 1$, and so that $a_{mt} = a_{tp} = 1$. Since $a_{tt} = 0$ and $a_{mt} = 1$ by Lemma 2.1.4, there exists k ε $\underline{n} \setminus \{t\}$ such that $a_{mk} = a_{kt} = 1$. But $a_{kt} = a_{tp} = 1$, which implies that $a_{kp} = 1$ since A ε Idem (B_n). Thus since $a_{mk}a_{kp}$ appears as a summand of c_{ij}, we conclude that $c_{ij} = 1$. This completes the proof. □

LEMMA 2.1.7. Let A ε Idem (B_n) be such that $a_{ii} = 1$ for all i. If A_{*t} is dependent on the other columns of A, then A_{t*} is dependent on the other rows of A. Of course a dual result holds.

We now sharpen up the statement of Lemma 2.1.7 below.

LEMMA 2.1.8. Let A ε Idem (B_n) be such that $a_{ii} = 1$ for all i. If A_{*t} is dependent on the other columns of A, then there exists s ε $\setminus \{t\}$ such that $A_{*t} = A_{*s}$ and $A_{t*} = A_{s*}$.

LEMMA 2.1.9. Let A ε Idem (B_n) be such that $a_{ii} = 1$ for all i. If A_{*t} is dependent on the other columns of A and if B ε B_{n-1} is a matrix obtained from A by removing the elements of A_{t*} and A_{*t}, then (i) $\rho_r(A) = \rho_r(B)$ and $\rho_c(A) = \rho_c(B)$, (ii) B ε Idem (B_n).

Proof. (i) Follows from Lemma 2.1.7.

(ii) Using the notation of Lemma 2.1.6 we see that if $a_{f(i),f(j)} = 1$, then $c_{ij} = 1$ since $a_{f(j),f(j)} = 1$ and $a_{f(i),f(j)}a_{f(j),f(j)}$ appears in the sum which makes up c_{ij}. This proves the lemma. □

THEOREM 2.1.10 [205, 244]. Let $A \in$ Idem (B_n). Then $\rho_r(A) = \rho_c(A) = \rho_s(A)$.

Proof. By symmetry it suffices to show that $\rho_r(A) = \rho_s(A)$. By Theorem 1.4.1, $\rho_s(A) \leq \rho_r(A)$. Let $\rho_r(A) = r$, $\rho_s(A) = s$ and write $A = A_1 + A_2 + \ldots + A_s$ where A_i has row rank 1. Choose a subset $\underline{i}_r = \{i_1, i_2, \ldots, i_r\}$ of $\underline{n}$ such that for $k \in \underline{i}_r$, A_{k*} forms a row basis of A and $a_{kk} = 1$ by Lemma 2.1.4. Suppose for some j, $k \in \underline{i}_r$ where $j \neq k$ both $a_{kj} = 1$ and $a_{jk} = 1$. Then by Remark 2.1.1, $A_{k*} \geq A_{j*}$ and $A_{j*} \geq A_{k*}$ so $A_{k*} = A_{j*}$. But since A_{k*} and A_{j*} were part of a row basis this is impossible. So for j, $k \in \underline{i}_r$ either $a_{kj} = 0$ or $a_{jk} = 0$. Thus no row rank 1 summand A_{i*} can have both a_{kk} and a_{jj} equal to 1 since row rank 1 implies $a_{kj} = 1$ and $a_{jk} = 1$ for this summand. Thus all the entries a_{kk} for $k \in \underline{i}_r$ must come from different row rank 1 summands. Thus $s \geq |\underline{i}_r| = r$. Thus $\rho_r(A) = \rho_s(A)$. This proves the theorem. □

COROLLARY 2.1.11. If $A \in$ Reg (B_n), then $\rho_r(A) = \rho_c(A) = \rho_s(A)$.

Proof. By Proposition 2.1.1, $A \in$ Reg (B_n) if and only if there exists an $B \in$ Idem (B_n) such that $B \in D_A$. But $\rho_r(B) = \rho_c(B)$ by Theorem 2.1.10 and so $\rho_r(A) = \rho_c(A)$ by Theorem 1.3.3. It follows from Definition 1.4.2 that the Schein rank of any two matrices in the same D-class is equal. This proves the corollary. □

The above corollary allows us to just talk about the rank of a regular matrix, and so from now on we will no longer distinguish between row and column rank and will denote both by ρ for regular matrices.

Our next series of results will be related to obtaining *basic idempotents*. These are idempotents in a D-class having the simplest possible form (see Theorem 2.1.13). The simplest idempotents in a given L-class are called *row reduced idempotents*, and these are studied next.

LEMMA 2.1.12 [244]. Let $A \in \text{Idem}(B_n)$. If $A[i|j]$ is a matrix obtained from A by interchanging the ith row with the jth row and the ith column with jth column, then $A[i|j] \in \text{Idem}(B_n)$ and $A[i|j] \mathcal{D} A$.

THEOREM 2.1.13 [244]. If $C \in \text{Reg}(B_n)$ and $\rho(C) = h$, then there exists $E \in B_n$ such that (i) $E \in D_C$, (ii) $E \in \text{Idem}(B_n)$ such that E has the form

$$\text{(2.1.1)} \qquad \left[\begin{array}{c|c} T_k & 0 \\ \hline 0 & 0 \end{array}\right], \qquad T_k = \begin{bmatrix} 1 & 0 & 0 & \cdots & 0 \\ * & 1 & 0 & \cdots & 0 \\ & & \cdots\cdots & & \\ * & * & * & \cdots & 1 \end{bmatrix}_{k\times k} \qquad (k \le n)$$

where $e_{ii} = 1$ for all $i \in \underline{h} \subset \underline{n}$ and $e_{ij} = 0$ if $i > h$ or if $j < i$.

Proof. We may assume that $C \neq 0$, since $\rho(0) = 0$, and we can let $E = 0$. Since C is regular there exists an $M \in \text{Idem}(D_C)$. Let F be the matrix obtained by repeating column and row interchanges of the kind described in Lemma 2.1.12, until we arrive at a matrix whose first h rows are exactly the elements of its row basis arranged in order of nondecreasing rank, i.e., we obtain a matrix $F \in D_C$ such that $F^2 = F$, $\rho(F) = h$, $\{F_{i*}\}_{i\in\underline{h}\subset\underline{n}} = B_r(F)$ and $w(F_{i*}) \le w(F_{j*})$ for all $i, j \in \underline{h}$ such that $j > i$. Such a matrix F exists and can be broken down into two operations: an interchange of rows followed by an interchange of columns. Since F is an idempotent and the first h rows of F comprise the basis elements it follows that $f_{ij} = 0$ for all i, j such that $1 \le i < j \le h$ since the first h rows are arranged in order of nondecreasing rank. We will show that F can actually be chosen such that $f_{ii} = 1$ for all $i \in \underline{h}$. We proceed as follows: if $f_{ii} \neq 1$ for some $i \in \underline{h}$, then since $F^2 = F$ and $F_{i*} \in B_r(A)$, there must exist $j \in \underline{n} \setminus \underline{h}$ such that $f_{ij} = 1$ and $F_{j*} = F_{i*}$. If we form $F[i|j]$ as in Lemma 2.1.12 we get an idempotent which has a 1 at its (i, i)-th entry. Thus inductively we may finally choose F so that it has all of its previous properties and such that $f_{ii} = 1$ for all $i \in \underline{h}$, i.e., F has the desired form (2.1.1).

Let $B \in B_n$ be such that

$$B_{i*} = \begin{cases} F_{i*} & \text{for all } i \in \underline{h} \\ 0 & \text{for all } i \notin \underline{h} \end{cases}$$

Thus $R(B) = R(F)$ and $B \in D_C$ and is of the same form as F except that all the rows below the kth are 0. Since $F^2 = F$ and $f_{ii} = 1$ for all $i \in \underline{h}$, $B^2 = B$. Note also that $w(F_{i*}) = w(F_{j*})$ for all $1 \le i \le j \le h$. Since $F_{*h} = e^h$, we must have that $F_{*h} \in B_c(B)$. We wish to alter B so that its first h columns form a basis for its column space and so that it still retains all of the properties which it had earlier. We proceed in a way similar to the way we had obtained F from M. We begin at the hth column and work toward the first column. Let $j \in \underline{h}$ be such that $B_{*j} \notin B_c(B)$, $e^j \le B_{*s} \le B_{*j}$. If we interchange B_{*j} and B_{*s}, we get an idempotent say K such that $K_{*j} \in B_c(K)$. Proceeding in this way inductively we finally get a matrix $Z \in D_C$ such that Z is an idempotent so that $z_{ii} = 1$ for $i \in \underline{h}$, and the first h rows of Z form the basis of the row space, while the first h columns form the basis of the column space. Thus Z is of the same form as F. Finally we let $E \in B_n$ such that $E_{*i} = Z_{*i}$ for all $i \in \underline{h}$ and $E_{*i} = 0$ for all $i \notin \underline{h}$. Since the first h columns of Z form the basis of its column space, we have that $C(E) = C(Z)$ and so $E \in D_C$. It follows that E is of the form illustrated in (2.1.1) of the statement of the theorem and is an idempotent since $e_{ii} = 1$, Z was an idempotent, and the fact that $Z_{i*} = 0$ for all $i \in \underline{h}$. Once E has been obtained, it is possible to order the rows so that the nonzero rows are monotone increasing with respect to rank. This proves the theorem. □

DEFINITION 2.1.4. Let $A \in \text{Idem}(B_n)$ having the properties listed in Theorem 2.1.13 then we say A is a *basic idempotent*.

The following definition is due to R. J. Plemmons and M. T. West [286].

DEFINITION 2.1.5. If $A \in B_{mn}$, then A is said to be *row (column) reduced* if the nonzero rows (columns) of A are an independent set. Now A is said to be *reduced* if it is both row and column reduced.

Note than a basic idempotent is reduced, although not every reduced idempotent is a basic idempotent. Thus Theorem 2.1.13 implies that every regular D-class of B_n contains at least one idempotent whi is reduced. If $A \in B_n$, then A is row (column) reduced if and only if A has k (m) nonzero rows (columns) where $\rho_r(A) = k$ ($\rho_c(A) = m$).

PROPOSITION 2.1.14 [286]. If $A \in \text{Reg}(B_n)$, then there exists at least one row (column) reduced idempotent in L_A (R_A).

PROPOSITION 2.1.15 [286]. If $A \in \text{Reg}(B_n)$, then there exists at most one reduced idempotent in L_A or R_A. Note that there may be no reduced idempotent in L_A and R_A.

We now present theorems which characterize regularity in terms of representability by the product of elements in the same D-class. These theorems will be very important when we discuss the problem of counting idempotents.

THEOREM 2.1.16 [244]. Let $G \in \text{Reg}(B_n)$. If $M \in D_G$, then there exist $A, B \in B_n$ such that (i) $AM \in R_G$, (ii) $MB \in L_G$, (iii) $(AM)(MB) = G$, if $M^2 = M$, (iv) $G^2 = G$ if and only if $(MB)(AM) = M$, (v) if (i)-(iv hold, then $AMB = G$ for any $A, B \in B_n$.

Proof. (i)-(iii) Let $M \in D_G$. There exist $A, B, C, D \in B_n$ such that $AMB = G$, $CAM = M$, $MBD = M$, $CGD = M$, $ACG = G$. Since $(AM)B = G$ and $GD = AM$, $AM \in R_G$ by Lemma 1.3.1. Likewise, $MB \in L_G$. Finally, $(AM)(MB) = AM^2B = AMB = G$.

(iv) We use the same symbols with the same meaning as above.

Necessity. We have (MB)(AM) = (CAMB)(AMBD) = (CG)(GD) = CG^2D = CGD = M.

Sufficiency. From necessity it follows that CG^2D = M. If CG^2D = M, we also have that G^2DB = G(GDB) = G^2 and similarly $ACG^2 = G^2$. Thus we see that AMB = G^2. But AMB = G and so G^2 = G.

(v) Clear. □

The following theorem shows that the representability of the type of Theorem 2.1.16 characterizes regular elements.

THEOREM 2.1.17 [244]. If $G \in B_n$, then $G \in \text{Reg}(B_n)$ if and only if there exist $X \in R_G$ and $Y \in L_G$ such that XY = G, i.e., if and only if $G \in R_G L_G$.

Proof. Necessity. Follows from Theorem 2.1.16 and Proposition 2.1.1.

Sufficiency. By Lemma 1.3.1 there exist A, B, C, $D \in B_n$ such that AY = G, CG = Y, XB = G, and GD = X. Let F = DC, then GFG = (GD) · (CG) = XY = G. This proves the theorem. □

In connection with the above two theorems it should be pointed out that these results hold for any semigroup.

REMARK 2.1.2. In general there is no uniqueness of representation as a product of elements in the same D-class for regular elements. For instance, if

$$G = \begin{bmatrix} 1 & 1 & 0 \\ 1 & 1 & 0 \\ 1 & 1 & 1 \end{bmatrix}, \quad Y = \begin{bmatrix} 1 & 1 & 0 \\ 1 & 1 & 1 \\ 0 & 0 & 0 \end{bmatrix}, \quad X = \begin{bmatrix} 1 & 0 & 0 \\ 1 & 0 & 0 \\ 1 & 1 & 0 \end{bmatrix}, \quad Z = \begin{bmatrix} 1 & 0 & 1 \\ 1 & 0 & 1 \\ 1 & 1 & 1 \end{bmatrix}$$

Then G^2 = G, i.e., $G \in \text{Reg}(B_3)$, $Y \in L_G$, X, $Z \in R_G$, and ZY = XY = G. However, if we put some more restrictions on X, Y we can have uniqueness of representation of regular matrices in product form.

In what follows we shall give quantitative results due to G. Markowsky [244, 245]. These results will depend heavily on the representation theorems, and particularly on the following lemma.

LEMMA 2.1.18 [244]. Let $A \in$ Reg (B_n). If $E \in D_A$ is a basic idempotent, then (i) for every L-class contained in D_A, there exists a $Y \in R_E$ such that Y is row reduced, and only the first h rows are nonzero, where $\rho(A) = h$, (ii) if for every L-class in D_A, say L_C, we pick a row reduced matrix $Y_{L_C} \in R_E$, then for any $G \in D_A$ there exists a unique $X \in R_G$ such that $XY_{L_G} = G$ and $X \in L_E$. Note that X is column reduced and that only its first h columns are nonzero, (iii) if $G \in D_A$ and $X \in R_G \cap L_E$ is such that $XY_{L_G} = G$, then $G \in$ Idem (B_n) if and only if $Y_{L_G}X = E$.

DEFINITION 2.1.6. Let E be a basic idempotent in D_A. Let v_1, $v_2, \ldots, v_t$ be all the vectors of C(E). Let $\alpha_i = \{p \in \underline{t}: e^p \le v_i\}$ for each $i \in \underline{t}$, and $\beta_i = |\{w \in R(E): w \le \bigwedge_{p\in\alpha_i} E_{p*}\}|$. (Note that $\beta_i \ge 1$ for all $i \in \underline{t}$ since $0 \in R(E)$). Finally let $m = \sum_{i=1}^{t} \beta_i$. Then we say m is the *Markowsky number*.

THEOREM 2.1.19 [244]. Let $A \in$ Reg (B_n) with $\rho(A) = r$. Then

$$|\text{Idem } (D_A)| = \sum_{i=0}^{r} \frac{(-1)^i \binom{r}{i} (m - i)^n}{|H_A|}$$

Proof. If S, T, U, $V \in B_n$, then S L T and U R V implies SU D TV. Thus in view of Lemma 2.1.18 and Remark 2.1.2, we will first attempt to count the number of pairs of elements X, $Y \in B_n$ such that $X \in L_E$, $Y \in R_E$ and YX = E. Since E is a basic idempotent, Y must be row reduced and X must be column reduced. The way we shall count the number of permissible pairs X, Y is by picking an element of C(E) to serve as the kth column of Y and then examining how our choice limits the number of elements of R(E) which can be serve as the kth row of X, i.e., we view the problem of constructing a pair of elements of B_n, say X, Y such that $X \in L_E$, $Y \in R_E$ and $YX = E$ in terms of picking pairs consisting of an element of C(E) to serve as the kth column of Y, and an element of R(E) to serve as the kth row of X. Thus if $e^p \le Y_{*k}$ for some $k \in \underline{n}$, we must have that $X_{k*} \le E_{p*}$ since YX = E. In particular, if $Y_{*k} = v_j$ for some j it follows that $X_{k*} \le \bigwedge_{p \in \alpha_j} E_{p*}$, i.e., β_j permissible choices for X_{k*} from the elements of R(E). Thus there are m different pairs of a kth column for Y and a kth row for X. The key to the whole matter is that any permissible pair X, Y must be formed from these m pairs and that the only condition which must be met in order that $X \in L_E$, $Y \in R_E$ and $YX = E$ is that exactly r (where $r = \rho(A)$) of these (column, row)) pairs appear as at least one out of the n pairs (repetition is allowed) which we use to construct X and Y. We will now proceed to prove the last statement, but we will break the proof into several steps.

(1) Any X and Y formed from the m permissible (column, row) pairs have the property that $C(Y) \subset C(E)$ and $R(X) \subset R(E)$. This statement is clearly true.

(2) If X and Y are formed in the manner described above from the m permissible (column, row) pairs then we must have that $(YX)_{s*} \le E_{s*}$ for all $s \in \underline{n}$. This is true since $e^s \le Y_{*k}$ implies $X_{k*} \le E_{s*}$ where $k \in \underline{n}$.

(3) For each $s \in \underline{h} \subset \underline{n}$, E_{*s} must appear at least once as a column of Y and E_{s*} must appear at least once as a row of X. Since we require that C(Y) = C(E) and R(E) = R(X), E is a basic idempotent, and because of Theorem 1.1.1 it follows that (3) is true.

(4) Let $w \in C(E)$ be such that $e^p \le w$, then $E_{*p} \le w$ if $p \in \underline{h} \subset \underline{n}$. This statement is true since E is an idempotent such that $e_{pp} = 1$, i.e., if $e^p \le E_{*s}$ for some $s \in \underline{n}$, then $E_{*p} \le E_{*s}$ and so the statement follows since the columns of E generate C(E).

(5) Let Y, X be formed from the m permissible (column, row) pairs. Then YX = E if and only if for each $s \in \underline{h} \subset \underline{n}$ there exists a $k \in \underline{n}$ such that $Y_{*k} = E_{*s}$ and $X_{k*} = E_{s*}$.

Sufficiency. By (2), $(YX)_{s*} \leq E_{s*}$. But $e_{ss} = 1$. So $y_{sk} = 1$. Thus $(YX)_{s*} \geq X_{k*} = E_{s*}$. So we must have that $(YX)_{s*} = E_{s*}$. Thus since Y is row reduced and X is column reduced it follows from (2) that $YX = E$.

Necessity. From Proposition 1.2.2 and (1) it follows that $C(Y) = C(E)$ and $R(X) = R(E)$. Since $YX = E$ and for each $s \in \underline{h}$, E_{s*} is an element of the basis of $R(E)$ it follows from (2) that there exists a $k \in \underline{n}$ such that $e^s \leq Y_{*k}$ and $X_{k*} = E_{s*}$. It follows from (4) that $Y_{*k} \geq E_{*s}$. If $Y_{*k} \neq E_{*s}$ there would exists a $p \in \underline{h}$ such that $y_{pk} = 1$ but $e_{ps} = 0$. Since $x_{ks} = e_{ss} = 1$ and since $YX = E$ we would have to have that $e_{ps} = 1$, but this would contradict our assumption that $e_{ps} = 0$. Thus $Y_{*k} = E_{*s}$. In view of the preceding remarks and (1)-(5), we can conclude that the number of pairs $Y \in R_E$ and $X \in L_E$ such that $YX = E$ is

$$\sum_{i=0}^{r} (-1)^i \binom{r}{i} (m - i)^n$$

by Lemma 1.3.5. Each such pair X, Y gives us an idempotent $XY \in D_A$. Lemma 2.1.18 tells us that each idempotent $F \in D_A$ has a unique representation providing we settle upon an element of $R_E \cap L_F$. But $R_E \cap L_F$ is an H-class with $|H_A|$ elements in it. It is clear that in our enumeration of pairs X, Y we will have $|H_A|$ representations of each idempotent in D_A. Thus the total number of idempotents in D_A is

$$\sum_{i=0}^{r} \frac{(-1)^i \binom{r}{i} (m - i)^n}{|H_A|}$$

This completes the proof of the theorem. □

COROLLARY 2.1.20. If A $\text{Reg}\,(B_n)$ is nonsingular, then there exist

$$\frac{n!}{|H_A|}$$

idempotents in D_A.

Proof. Follows clearly from Theorem 2.1.19 and Corollary 1.3.6. □

Note that it is not too hard to calculate Markowsky number m for some D-classes and we will give an example below.

EXAMPLE 2.1.1. Let $I_{nr} = (\delta_{ij}) \in B_n$ be such that $\delta_{ij} = 1$ if $i = j \in \underline{h} \subset \underline{n}$ and $\delta_{ij} = 0$ otherwise. Then clearly I_{nr} is a basic idempotent with $B_c(I_{nr}) = \{e^1, e^2, \ldots, e^r\}$ and $B_r(I_{nr}) = \{e_1, e_2, \ldots, e_r\}$. First we have the r permissible pairs (e^i, e_i) for every $i \in \underline{h}$. Next we have the 2^r pairs (v, 0) where $v \in C(I_{nr})$ and $0 \in R(I_{nr})$. Finally, we have the $2^r - 1$ pairs (0, v) where $0 \in C(I_{nr})$ and $v \in R(I_{nr}) \setminus \{0\}$, since we do not want to count (0, 0) twice. It is clear that these are the only possible pairs and that thus $m = 2^r + 2^r - 1 + r = 2^{r+1} + r - 1$.

Example 2.1.1 illustrates the fact that often the calculation of m is made easier by determining first the h elements which must always appear among the elements which are used to make up X and Y, since often a pattern emerges in the remaining cases which simplifies the calculation of m - r.

We next show that Theorem 2.1.20 can be formulated in terms of antichains. Recall (Definition 1.4.4) that an antichain in a partially ordered set S is a subset X of S such that for no $x, y \in X$ is $x < y$.

LEMMA 2.1.21 [22]. For any finite distributive lattice L, let B be a basis for L. We regard B as a poset with ordering induced by L. The mapping from the antichains on B to L sending an antichain to the sum of its elements (and ∅ to zero) is an isomorphism.

Proof. Any element of L is a sum of some set of basis elements. By taking only the maximal elements of the sum we obtain an antichain with the same sum. Suppose two antichains X and Y have the same sum.

Let a be a maximal element of the symmetric difference $X \Delta Y$. Suppose $a \in X$. Then a is not less than or equal to any element b of Y, or b would have to lie in $X \cap Y$, by maximality of a. If $b \in X \cap Y$, $a \in X \Delta Y$ then $a \neq b$. And $a < b$ would imply X is not an antichain.

Multiply both sides of the equation

$$\sum_{x \in X} x = \sum_{y \in Y} y$$

by a and use distributivity. Then

$$a = \sum_{y \in Y} ya.$$

Each ya is less than a since a is not less than or equal to any element of Y. But a is a basis element so it cannot be expressed as a sum of smaller elements of L.

Thus no two different antichains have the same image. This proves the lemma. □

DEFINITION 2.1.7. Let E be a basic idempotent. Then m(E) is the number of ordered pairs A, B of antichains in $B_r(E)$ such that every element of A is less than or equal to every element of B.

THEOREM 2.1.22 [206]. The Markowsky number m occurring in Theorem 2.1.20 is equal to m(E).

Proof. Suppose v is the sum of the basis vectors in some antichain A(v). By Lemma 2.1.21, A(v) is uniquely determined. Then $p \in \alpha_i$ if and only if $v_p = 1$ if and only if $(E_{*j})_p = 1$ for some E_{*j} in the antichain A(v). And $(E_{*j})_p = 1$ means $e_{pj} = 1$. By Remark 2.1.1 this is so if and only if $E_{*j} \geq E_{*p}$.

Let f be the mapping $B_r(E) \to B_c(E)$ sending $E_{j*} \to E_{*j}$. Then f is one-to-one and onto and reverses all poset inequalities. Thus f gives a one-to-one correspondence on antichains.

Thus $p \in \alpha_i$ if and only if $E_{p*} \geq E_{j*}$ for some $E_{j*} \in f(A(v))$. And $w \leq \bigwedge_{p \in \alpha_i} E_{p*}$ if and only if w is less than or equal to all E_{p*}

such that E_{p*} is greater than or equal to some $E_{j*} \varepsilon f(A(v))$ if and only if w is less than or equal to all $E_{j*} \varepsilon f(A(v))$.

Furthermore $\{w \varepsilon R(E): w \leq \bigwedge_{p\varepsilon\alpha_i} E_{p*}\}$ is the subsemigroup generated by all basis vectors w_0 such that w_0 is less than or equal to all $E_{j*} \varepsilon f(A(v))$.

This subsemigroup is in fact a sublattice and so distributive. By Lemma 2.1.21 it is isomorphic to the set of all antichains B such that every element of B is less than or equal to every element of $f(A(v))$. Thus $\beta_i = |\{\text{antichains } B: \text{for all } b \varepsilon B, \text{ for all } a \varepsilon f(A(v)), b \leq a\}|$. As we sum β_i over v, $f(A(v))$ will range once over every antichain in $B_r(E)$. Therefore $m = \Sigma \beta_i = |\{(A, B): A, B \varepsilon \zeta, a \leq b$ for all $a \varepsilon A, b \varepsilon B\}|$ where ζ is the set of antichains of $B_r(E)$. This proves the theorem. □

EXAMPLE 2.1.2. Let

$$E = \begin{bmatrix} 1 & 1 & 1 \\ 0 & 1 & 0 \\ 0 & 0 & 1 \end{bmatrix}$$

Then P is the poset {1, 2, 3} with partial order 1 > 2, 1 > 3 and no other strict inequalities. The antichains of P are ∅, {1}, {2}, {3}, {2, 3}. The respective numbers of antichains each of whose elements are less than or equal to all the elements of one of these is 5, 5, 2, 2, 1. Thus $m(E) = 15$.

From the present definition of m(E) it is possible to derive a formula for $m(E \oplus F)$, a formula for matrices such as

$$\begin{bmatrix} E & 0 \\ J & F \end{bmatrix}$$

and upper and lower bounds for m(E) in terms of $\rho(E)$ and $|R(E)|$. Here $E \oplus F$ denotes the direct sum of E and F.

PROPOSITION 2.1.23. Let E be a basic idempotent matrix. Then $m(E) \geq 2|R(E)| + \rho(E) - 1$.

Proof. Count pairs of antichains where $A = \emptyset$ or $B = \emptyset$ or $A = B = \{x\}$. □

We next give a criterion of idempotency due to B. M. Schein [333]. We need the following definitions.

DEFINITION 2.1.8. Let $A \in B_n$. If A is said to be *reflexive* then $a_{ii} = 1$ for every i. We say A is *transitive* if $a_{ij} = 1$ and $a_{jk} = 1$ then $a_{ik} = 1$. We say A is a *quasi-order* if A is reflexive and transitive. (Dually, a *quasi-order relation* is a relation which is reflexive and transitive).

The matrix of any quasi-order is idempotent, however some idempotents are not reflexive. Likewise all idempotents are transitive in the sense that $A^2 \le A$, but some transitive matrices are not idempotent. Thus, quasi-order matrices $\subset$ idempotent matrices $\subset$ transitive matrices where each inclusion is proper.

DEFINITION 2.1.9. A subset $\underline{m}$ and $\underline{n}$ is *permissible* for a matrix $A \in B_n$ if and only if (1) for $i \in \underline{m}$, $j \in \underline{n}$, $i \ne j$ it is not true that $a_{ij} = a_{ji} = 1$, (2) for $i, j \in \underline{m}$, $i \ne j$, if $a_{ij} = 1$ then there exists $h \in \underline{n}$, $h \ne i$, $h \ne j$ such that $a_{ih} = a_{hj} = 1$.

THEOREM 2.1.24 [333]. $B \in \text{Idem}(B_n)$ if and only if there exists a quasi-order matrix A and a permissible subset $\underline{m}$ for A such that $b_{ij} = 0$ if $i = j$, $i \in \underline{m}$, $b_{ij} = a_{ij}$ otherwise.

Proof. Let A be a quasi-order matrix and $\underline{m}$ a permissible subset of $\underline{n}$ and let B be defined by the equation above. Then $B \le A$ so $B^2 \le A^2 = A$. For $i \in \underline{m}$, $x \in \underline{n}$, $b_{ix}b_{xi} = 0$ by the first condition. Thus $b_{ii} = 0$ for $i \in \underline{m}$. Thus $B^2 \le B$.

Suppose $b_{ij} = 1$, i or j not in $\underline{m}$. Then $b_{ii} = 1$ or $b_{jj} = 1$ so that $b^{(2)}_{ij} = 1$. Suppose $b_{ij} = 1$ for $i, j \in \underline{m}$. Then $i \ne j$ by definition of B. By the second condition there exists h such that $a_{ih} = a_{hj} = 1$,

$h \neq i$, $h \neq j$. Then by definition of B, $b_{ih} = b_{hj} = 1$. Thus $b_{ij}^{(2)} = 1$. This proves $B^2 \geq B$. So B is idempotent.

Now let $B \in \text{Idem}(B_n)$. Let $A = B + I$. Then $A^2 = A$ and A is reflexive. Let $\underline{m} = \{i \in \underline{n}: b_{ii} = 0\}$. Suppose there existed $i \in \underline{m}$, $j \in \underline{n}$, $i \neq j$ such that $a_{ij} = a_{ji} = 1$. Then $b_{ij} = b_{ji} = 1$ since the off-main diagonal elements of B are the same as those of A. So $b_{ii}^{(2)} = 1$. Thus $b_{ii} = 1$ since $B^2 = B$. But this contradicts the definition of $\underline{m}$. So $\underline{m}$ satisfies the first condition.

Suppose $i, j \in \underline{m}$, $i \neq j$, $a_{ij} = 1$. Then $b_{ij} = 1$. Thus $b_{ij}^{(2)} = 1$. So there exists h such that $b_{ih} = b_{hj} = 1$. If $h = i$ or $h = j$ then $h \in \underline{m}$, $b_{hh} = 1$ which contradicts the definition of $\underline{m}$. So $h \neq i$, $h \neq j$. Thus $\underline{m}$ satisfies the second condition.

It follows from the definitions of A, $\underline{m}$ that B satisfies the equations of the theorem. This proves the theorem. □

We now give a computational method of checking whether a given matrix is regular. The following results are due to B. M. Schein [336].

DEFINITION 2.1.10. Let A and B be the Boolean matrices. If $ABA \leq A$ we call B a *subinverse* of A.

The set of subinverses of A is closed under addition, so there exists a largest subinverse $\overline{A}$, and B is a subinverse of A if and only if $B \leq \overline{A}$.

THEOREM 2.1.25 [336]. Let A be a Boolean matrix. Then $\overline{A} = (A^T A^C A^T)^C$. Here A^C denotes the complement of A, i.e., the matrix obtained from A by replacing all 0 entries by 1 and all 1 entries by 0.

Proof. Let E(x, y) denote the matrix which has a 1 in location (x, y) and zeros elsewhere. Then by the above remark, $\overline{a}_{xy} = 1$ if and only if E(x, y) is a subinverse of A. Also E(x, y) is a subinverse of A if and only if $AE(x, y)A \le A$, i.e., if for all u, v, $a_{ux}a_{yv} \le a_{uv}$. Thus $\bar{a}_{xy} = 0$ if and only if this is false. So $\bar{a}_{xy} = 0$ if and only if for some x, y, $a_{ux}a_{yv} > a_{uv}$, i.e., $a_{uv} = 0$, $a_{ux} = 1$, $a_{yv} = 1$. Then $\bar{a}_{xy} = 0$ if and only if $(a_{uv})^C = 1$, $a_{ux} = 1$, $a_{yv} = 1$ for some u, v. This holds if and only if $(A^T A^C A^T)_{xy} = 1$. Thus $\bar{a}_{xy} = 0$ if and only if $(A^T A^C A^T)_{xy} = 1$. Thus $\bar{A} = (A^T A^C A^T)^C$. This proves the theorem. □

THEOREM 2.1.26 [336]. A Boolean matrix A is regular if and only if $A \le A(A^T A^C A^T)^C A$.

Proof. Suppose this condition holds. It says $A \le A\bar{A}A$. But by definition of $\bar{A}$, $A\bar{A}A \le A$. So $A = A\bar{A}A$ and A is regular.

Suppose A is regular. Then $A = ABA$ for some B. But $B \le \bar{A}$ since B is a subinverse and $\bar{A}$ is the largest subinverse. So $A \le A\bar{A}A$. Thus $A = A\bar{A}A$. This proves the theorem. □

DEFINITION 2.1.11. Let A be a Boolean matrix. We say A_{i*} and A_{*j} *agree* if for all u, v, $a_{iu} = a_{vj} = 1$ implies $a_{uv} = 1$. The *agreement matrix* for A is the matrix B such that $b_{ij} = 1$ if and only if the ith row and jth column of A agree.

Theorem 2.1.25 can be interpreted as, the largest subinverse of A is the transpose of the agreement matrix of A.

We next state a theorem of E. S. Wolk [403] using this characterization of regularity. We need the following definition.

DEFINITION 2.1.12. A matrix $A \in B_n$ is called *antisymmetric* if $a_{ij} = 1$ then $a_{ji} = 0$.

THEOREM 2.1.27 [403]. Let A be a reflexive and antisymmetric Boolean matrix. Then A is regular if and only if it is transitive.

We proceed to establish a relationship between regularity and distributive lattices. We shall be making extensive use of the following definition and lemma which are due to K. H. Kim and F. W. Roush [194].

DEFINITION 2.1.13. For each $v \in B_r(A)$, a vector u with one 1 is called an *identification vector* if and only if $u \leq w$ holds if and only if $v \leq w$ for $w \in B_r(A)$. Such a vector will be denoted by I(v).

LEMMA 2.1.28 [197]. If R(A) is distributive, then for every $v \in B_r(A)$ there exists at least one $u \in I(v)$.

Proof. A vector u containing only one 1 is an identification vector v if and only if $u \leq v$ and $u \nleq \sum_{w \in S} w$ where S is the set of row basis vectors which are not greater than or equal to v. Thus there exists at least one identification vector v if and only if $v \nleq \sum_{w \in S} w$.

Suppose on the contrary $v \leq \sum_{w \in S} w$. Multiply both sides by v, in the row space regarded as a lattice. Thus $v = \left(\sum_{w \in S} w\right)v$. By distributivity, $v = \sum_{w \in S} wv$. Each vector wv is properly less than v, since each w was not greater than or equal to v. Thus v is a sum of properly smaller vectors, this implies that v is not a row basis vector, which is a contradiction. This proves the lemma. □

DEFINITION 2.1.14. A matrix $A \in B_n$ is called *nonsingular* if $A \in \text{Reg}(B_n)$ and $\rho(A) = n$.

THEOREM 2.1.29 [412]. A finite Boolean matrix is regular if and only if its row space is a distributive lattice.

Sketch of Proof. The proof of this result can readily be obtained by combining proofs given at different points in this text.

As an intermediate step we show both (Reg) regularity and (Dis) distributivity of row space are equivalent to (Iden) there exists an idenfication vector for each row basis vector.

(1) (Dis) implies (Iden). This is the proof of Lemma 2.1.28.

(2) (Iden) implies (Reg). This is shown by our algorithm for a generalized inverse in Chapter 3.

(3) (Reg) implies (Dis). Let A be regular. Then there exists an idempotent E in the same D-class as A, such that E is direct sum of zero and a nonsingular idempotent F by Theorem 2.1.13. Thus $R(A) \simeq R(E) \simeq R(F)$. The matrix F is the matrix of partial order relation ρ in which $f_{ij} = 1$ if and only if $i \leq j$ in ρ. This implies that for any i, j the vector $F_{i*} \cdot F_{j*}$ is a union of rows of F. Thus R(F) is closed under intersection. This implies R(F) is a sublattice of V_r where r is the rank of A. Since V_r is distributive, R(F) will also be distributive. □

2.2 Maximal Subgroups

Earlier it has been said that B_n is not a group and in fact the only invertible elements of B_n are permutation matrices. The permutation matrices P_n form a subgroup of the semigroup B_n. However they are not the only subsemigroup of B_n which forms a group. By Green's theorem every H-class containing an idempotent is a (maximal) subgroup of B_n. In this section we study the group structure of these subgroups. It turns out that any maximal subgroup of B_n is isomorphic to a subgroup of the symmetric group on n symbols and that every finite group is isomorphic to some maximal subgroup of B_n.

In 1969 J. S. Montague and R. J. Plemmons [257] proved that every finite group is a maximal subgroup of B_n for large n. This was extended by R. J. Plemmons and B. M. Schein [285] to the general case, using theorem due to K. A. Zaretski [412]. A. H. Clifford [54] gave an entirely self-contained proof of this interesting theorem. This is in contrast with the fact that all maximal subgroups of the semigroup of partial transformations are symmetric groups.

THEOREM 2.2.1 [257]. A group is a maximal subgroup of B_n if and only if it is the group of automorphisms of some poset of size less than or equal to n.

Proof. Suppose G is a maximal subgroup of B_n. Then G is an H-class of B_n containing an idempotent E. By a slight extension of Theorem 1.3.4, G is isomorphic to Aut (R(E)), where Aut (R(E)) denotes the automorphism of R(E). Thus G depends only on the D-class of E. Thus we may assume E is a reduced idempotent, i.e., $E = E_1 \oplus 0$ where E_1 is a nonsingular idempotent. Here $E_1 \oplus 0$ denotes the direct sum of E_1 and 0. By Theorem 1.3.9, $G = \{P \varepsilon P_n: PE_1Q = E_1$ for some $Q \varepsilon P_n\}$. Recall Definition 1.2.7 that P_n denotes the set of all $n \times n$ permutation matrices. But the identity is the unique permutation less than E_1 since E_1 can be put into a triangular form.

Thus $PIQ = I$ so $Q = P^{-1}$ where P^{-1} denotes the inverse of P. Thus $G = \{P \varepsilon P_n: PE_1P^{-1} = E_1\}$. However this is precisely the automorphism group of the poset represented by E_1.

Conversely, given a poset whose automorphism group is G, we form its matrix E_1. Then $G \simeq \{P \varepsilon P_n: PE_1Q = E_1$ for some $Q \varepsilon P_n\} \simeq$ H-class of $E_1 \oplus 0$. This proves the theorem. □

G. Birkhoff [23] proved in 1946 that every finite group is the automorphism group of some finite poset. Thus we have the Montague-Plemmons-Schein-Clifford theorem as corollary.

Every maximal subgroup of B_n has a natural representation as a permutation group, i.e., permutations of a basis for R(E). Thus C.-Y. Chao [49] has considered the question, which permutation groups G can occur as maximal subgroups of B_n. By similar arguments to Theorem 2.2.1, a permutation group can occur if and only if it is the group of automorphisms of a poset of order n. The following results are a modified version of Chao's results.

LEMMA 2.2.2. Let h be an automorphism of a finite poset P and let $x > y$. Then $h(x) = y$ is impossible.

Proof. Suppose $h(x) = y$. Thus $h(x) < x$. Then $h(h(x)) < h(x)$, $h^3(x) < h^2(x)$, etc., and we have an infinite descending chain in the poset, which is impossible. □

PROPOSITION 2.2.3. If Group G acts on a finite poset, the orbits of G are antichains.

Proof. This follows from Lemma 2.2.2. □

PROPOSITION 2.2.4. If group G acts on a finite poset P, and x, w belong to the same orbit and y, z belong to the same orbit and $x > y$ then it is impossible that $z > w$.

PROPOSITION 2.2.5. If group G acts on a finite poset P, the orbits of G form a poset under the relation $O_1 > O_2$ if $x > y$ for some $x \in O_1$ and some $y \in O_2$.

DEFINITION 2.2.1. The *direct sum of a collection of permutations* is the direct product of the abstract groups acting on the disjoint union of the sets on which the group act, in the natural way.

THEOREM 2.2.6 [49]. If G_i are automorphism groups of posets for $i = 1$ to k, so is their direct sum.

Proof. Let G_i act on posets P_i. Let ω be the ordinal sum, in some order of the poset P_i. Then the direct sum of the G_i is the automorphism group of ω. □

THEOREM 2.2.7 [49]. Suppose G is the automorphism group of a poset P, and any two orbits of G have relatively prime sizes. Then G is a direct sum of symmetric groups.

Proof. Let O_1 and O_2 be two orbits of G. Then G is transitive on each of O_1 and O_2. We will show G is transitive on $O_1 \times O_2$.

Let H_1 be the isotropy group of an element x_1 of O_1 and H_2 the isotropy group of an element x_2 of O_2. Then $[G : H_1] = |O_1|$ and $[G : H_2] = |O_2|$, where $[G : H] = |G|/|H|$. The isotropy group of (x_1, x_2) is $H_1 \cap H_2$. Consider $[G : H_1 \cap H_2]$. It is $[G : H_1][H_1 : H_1 \cap H_2]$ and it is $[G : H_2][H_2 : H_1 \cap H_2]$. Thus $|O_1|$ and $|O_2|$ divide $[G : H_1 \cap H_2]$. Since $|O_1|$ and $|O_2|$ are relatively prime, $|O_1||O_2|$ divides $[G : H_1 \cap H_2]$. Thus the orbit of (x_1, x_2) has at least $|O_1||O_2|$ elements. Thus it is all of $O_1 \times O_2$. Thus G is transitive on $O_1 \times O_2$. This implies that if $x > y$ for some $x \in O_1$ and some $y \in O_2$ then $x > y$ for any $x \in O_1$ and any $y \in O_2$. This implies the partial order is invariant under any permutation which sends each orbit to itself. Thus G is the sum of the symmetric groups on the different orbits. This proves the theorem. □

We next consider more generally the case of groups having only two orbits. These results are due to K. H. Kim and F. W. Roush [208].

PROPOSITION 2.2.8 [208]. A group G with only two orbits of sizes r, s is the automorphism group of some poset if and only if G is the automorphism group $\{(P, Q): PXQ = X$, P and Q are permutation matrices$\}$ of some $r \times s$ matrix X, which has at least one 1 in each row and column.

Proof. Given a matrix X, take the poset

$$\begin{bmatrix} I_s & 0 \\ X & I_r \end{bmatrix}$$

Then $\{(P, Q): PXQ = X$, P and Q are permutation matrices$\}$ will be the automorphism group of X.

Given a group G both orbits will be antichains. Thus some element of one orbit must be greater than some element of the other orbit, or G would be the symmetric group and have only one orbit. Let $x_{ij} = 1$ if and only if the element i of the first orbit is comparable to the element j of the second orbit. Then X has at least one 1 in each row and column since G is separately transitive on the two orbits and so G must be the automorphism group of X. □

THEOREM 2.2.9 [208]. Let G have only two orbits, both of size p, where p is prime. Then G is the automorphism group of a poset if and only if G is the automorphism group of a nonzero $p \times p$ circulant X.

Proof. Sufficiency follows from the previous theorem. Suppose G is the automorphism group of some $p \times p$ matrix X with at least one 1 in each row and column. If X = J then $G = S_p \times S_p$, where S_p denotes the symmetric group on n symbols. So assume $G \neq S_p \times S_p$. Then G acts transitively on each orbit, so $[G : H_1] = p$ if H_1 is the isotropy group of an element in the first orbit. Thus a Sylow p-subgroup of G is larger than one of H_1. Let Y be an element of order p^s lying in G but not in H_1. Since $G \subset S_p \times S_p$, Y has order p. And on the first orbit, Y must act as a cycle. Since Y has order p, Y must act either as a cycle or as the identity on the second orbit. Suppose Y is the identity on the second orbit. Then each column of X is all zeros or all ones since if $Y = (P_0, I)$ then $P_0X = X$ where P_0 is a p-cycle. But since G acts transitively on the second orbit also, X must be all zeros or all ones. Thus X = 0 or X = J, but this is contrary to assumption. Thus $Y = P_0, P_1$ where P_0, P_1 are p-cycles. Let Z denote the p-cycle

$$\begin{bmatrix} 0 & 1 & 0 & \cdots & 0 \\ 0 & 0 & 1 & \cdots & 0 \\ 0 & 0 & 0 & \cdots & 0 \\ & & \cdots\cdots & & \\ 1 & 0 & 0 & \cdots & 0 \end{bmatrix}$$

Let $P_0 = RZR^{-1}$ and $P_1 = SZ^{-1}S^{-1}$. Let $X_1 = R^{-1}XS$. Here R^{-1} denotes the inverse of R. Then X_1 will have an automorphism group isomorphic to that of X, and $ZX_1Z^{-1} = X_1$. Thus X_1 is a circulant. This proves the theorem. □

2.3 Group-Complexity

Group-complexity is a new technique in semigroup theory which has its origins in the theory of automata. A semigroup is called *combinatorial* it its H-classes contain one element each. The *group-complexity* is more or less the number of stages consisting of combinatorial semigroups and groups required to construct the given semigroup. The material in this section is not needed to read the rest of the book.

The results in this section are due to J. Rhodes [303]. See also K. H. Kim and F. W. Roush [194].

DEFINITION 2.3.1. Let S_1 and S_2 be semigroups and Y a homomorphism from S_1 into the endomorphism semigroup of S_2. Then $S_2 \times_Y S_1$ (*semidirect product* of S_2 by S_1) is the set $S_2 \times S_1$ with product $(a, b)(c, d) = (aY(b)(c), bd)$.

For simplicity $S_n \times_{Y_{n-1}} \cdots \times_{Y_2} S_2 \times_{Y_1} S_1$ denotes the iterated semidirect product $((\ldots (S_n \times_{Y_{n-1}} S_{n-1}) \times_{Y_{n-2}} S_{n-2}) \ldots \times_{Y_1} S_1)$.

DEFINITION 2.3.2. The *group-complexity* $\#_G(S)$ of a finite semigroup S is the least nonnegative integer n such that S is a homomorphic image of a subsemigroup of some semigroup $C_n \times_{Y_{n-1}} G_n \times_{Z_{n-1}} C_{n-1} \times_{Y_{n-2}} G_{n-1} \times_{Z_{n-2}} \cdots C_1 \times_{Y_0} G_1 \times_{Z_0} C_0$ where C_i are finite semigroups whose H-classes contain one element each, and the G_i are finite groups.

J. Rhodes [303] has proved the following:

AXIOM. If S is a subdirect product of $S_1, S_2, \ldots, S_n$ then $\#_G(S) = \max \{\#_G(S_i)\}$.

(R1) Suppose S has a unique maximal J-class which is regular. Let e be any idempotent of J-class. Then $\#_G(S) = \#_G(eSe)$.

(R2) Let K be an ideal of S. Then $\#_G(S) \le \#_G(S/K) + \#_G(K)$.

(R3) If every R-class of S contains at most one idempotent, then $\#_G(S) \le 1$.

(R4) Let S be the semigroup of transformations on an n-element set. Then $\#_G(S) = n - 1$.

For details, see J. Rhodes [304-308], J. Rhodes and B. Tilson [309], D. Allen [8], P. Stiffler [376], and B. Tilson [384].

It follows directly from the definition that if $T \subset S$ then $\#_G(T) \leq \#_G(S)$. Thus by (R4), $\#_G(B_n) \geq n - 1$. And it is true that $\#_G(B_1) = 0$ since B_1 has no H-classes containing more than one element.

DEFINITION 2.3.3. Let $\rho_s^{n-1}(B_n)$ denote the set of matrices of B_n which have Schein rank less than or equal to n - 1.

LEMMA 2.3.1. The set $\rho_s^{n-1}(B_n)$ is an ideal.

Proof. Let $X \in \rho_s^{n-1}(B_n)$. Then $X = X_1 + \ldots + X_{n-1}$ where X_i have rank 1 or rank 0. Thus $AXB = AX_1B + \ldots + AX_{n-1}B$. And the AX_iB will have rank 1 or rank 0. Thus $AXB \in \rho_s^{n-1}(B_n)$. □

LEMMA 2.3.2. Every regular matrix outside $\rho_s^{n-1}(B_n)$ has row and column rank n.

Proof. For regular matrices, $\rho_r = \rho_c = \rho_s$. □

LEMMA 2.3.3. An R-class in $B_n/\rho_s^{n-1}(B_n)$ has at most one idempotent.

Proof. This is equivalent to saying an R-class in $B_n \setminus \rho_s^{n-1}(B_n)$ has at most one idempotent since the mapping $B_n \setminus \rho_s^{n-1}(B_n)$ to $B_n/\rho_s^{n-1}(B_n)$ is one-to-one and is an isomorphism of sets on each R-class of $B_n/\rho_s^{n-1}(B_n)$.

Suppose R is an R-class of $B_n \setminus \rho_s^{n-1}(B_n)$ with two distinct idempotents E_1, E_2. Any idempotent of Schein rank n has row and column rank n and so can be put into a form:

$$\begin{bmatrix} 1 & 0 & 0 & \cdots & 0 \\ * & 1 & 0 & \cdots & 0 \\ * & * & 1 & \cdots & 0 \\ & & \cdots\cdots & & \\ * & * & * & \cdots & 1 \end{bmatrix}$$

by conjugation. Thus the identity is the unique permutation less than or equal to any rank n idempotent.

But since $E_1 \; R \; E_2$ the columns of E_1 are a permutation of those of E_2 since all columns are basis elements. Thus $E_1 = E_2Q$ for some permutation matrix Q. Thus $IQ \leq E_1$. Thus $IQ = I$. It follows that $Q = I$ and $E_1 = E_2$. This contradiction proves the lemma. □

LEMMA 2.3.4. The ideal $\rho_s^{n-1}(B_n)$ has a unique maximal J-class which is the J-class of the idempotent $E = I_{n-1} \oplus 0$. Furthermore $EB_nE = B_{n-1}$.

Proof. Exercise.

THEOREM 2.3.5 [303]. $\#_G(B_n) = n - 1$.

Proof. It was noted above that $\#_G(B_n) \geq n - 1$. By (R2),

$$\#_G(B_n) \leq \#_G(B_n/\rho_s^{n-1}(B_n) + \#_G(\rho_s^{n-1}(B_n))$$

By (R3) and Lemma 2.3.3, $\#_G(B_n/\rho_s^{n-1}(B_n)) \leq 1$. By (R1) and Lemma 2.3.4,

$$\#_G(\rho_s^{n-1}(B_n)) = \#_G(E\rho_s^{n-1}(B_n)E) = \#_G(B_{n-1})$$

Thus

$$\#_G(B_n) \leq 1 + \#_G(B_{n-1})$$

By induction, $\#_G(B_n) \leq n - 1$. This proves the theorem. □

EXAMPLE 2.3.1. The first two matrices either have row rank less than n or column rank less than n, but the last does not. Thus the set of matrices having row or column rank less than n is not an ideal.

For instance

$$\begin{bmatrix} 1 & 0 & 0 & 0 \\ 1 & 1 & 0 & 0 \\ 0 & 1 & 1 & 0 \\ 0 & 0 & 1 & 0 \end{bmatrix} \begin{bmatrix} 1 & 1 & 0 & 0 \\ 0 & 1 & 1 & 0 \\ 0 & 0 & 1 & 1 \\ 0 & 0 & 0 & 0 \end{bmatrix} = \begin{bmatrix} 1 & 1 & 0 & 0 \\ 1 & 1 & 1 & 0 \\ 0 & 1 & 1 & 1 \\ 0 & 0 & 1 & 1 \end{bmatrix}$$

2.4 Other Properties

The *maximal submonoids* $h_n(M)$ of B_n (the largest subsemigroups containing a given idempotent M) are the sets MB_nM. They were first considered by Z. Hedrlin (see B. M. Schein [339]). K. A. Zaretski [411] characterized them as translational hulls of matrix 0-bands. D. J. Hartfiel and C. J. Maxson [150] proved $h_n(M)$ is isomorphic to the semigroup of additive endomorphisms of R(M). It was proved in K. H. Kim and F. W. Roush [199] that $h_n(M)$ is a distributive lattice, having basis the matrices v^Tw where $v^T \in B_c(M)$ and $w \in B_r(M)$, that meet is given by $M(A \cap B)M$, that $h_n(M)$ is a regular semigroup if and only if R(M) is a chain of diamonds (an ordinal sum of posets 1 and 1 + 1), that $h_n(M)$ is isomorphic to the semigroup of monotone binary relations on the partially ordered set $B_r(M)$, and to the antichains in the poset $B_r(M) \times B_r^*(M)$, where $B_r^*(M)$ is the dual of $B_r(M)$, and that the maximal subgroups of $h_n(M)$ are the automorphism groups of all posets which are isotone images of subposets of $B_r(M)$. D. W. Hardy and M. C. Thornton [143-145] have also done work in this area.

In K. H. Kim and F. W. Roush [187] endomorphisms of B_n are classified in terms of their restrictions to regular elements. It is shown how the poset of D-classes determines the lattice of ideals, that every nontrivial semigroup congruence on B_n sends all rank 1 matrices to 0, and that every semigroup congruence on B_n which induces a nontrivial quotient of the symmetric group sends all regular D-classes to 0, except the symmetric group itself. Because of the vast number of irregular D-classes, to classify endomorphisms on them seems hopeless. An endomorphism other than an automorphism must send all regular elements of rank less than n to 0 (this is a difficult proof). The image of the matrix X such that X is reflexive and the only off-main diagonal

1 of X is x_{12} determines the image on regular elements (assuming we have identity on the symmetric group).

DEFINITION 2.4.1. The *permanent* of an $n \times n$ matrix A is the number Per (A) = $\Sigma\, a_{1\pi(1)} a_{2\pi(2)} \cdots a_{n\pi(n)}$ where the summation is carried out over all permutations π.

There is an endomorphism h of B_n such that h(A) = J if Per (A) $\neq$ 1 (over $\mathcal{R}$) and if Per (A) = 1, and A is reflexive, h(A) = M such that $m_{ij} = 1$ if and only if i = j or a_{ij} or a_{kj} or $a_{ik} = 1$ for some $k \neq i, j$ and h(PA) = P(h(A)) if P is a permutation matrix.

DEFINITION 2.4.2. A *Hall matrix* is an $n \times n$ Boolean matrix A such that $A \geq P$ for some $P \in P_n$. (These matrices are briefly discussed in Section 4.5 also).

In K. H. Kim and F. W. Roush [198] it is shown that the least possible degree of a faithful representation over a field of B_n, and the subsemigroups of Hall matrices and reflexive matrices are respectively $2^n - 1$, $2^n - 2$, $2^n - 2$. In the case of B_n it suffices to show that the semigroup ring of the rank 1 matrices with zero identified to the zero of the ring is isomorphic to the ring of all $(2^n - 1) \times (2^n - 1)$ matrices over a field.

In K. H. Kim and F. W. Roush [196] it is shown that the all regular Boolean matrices of B_n are generated by four matrices: two permutation matrices, a rank (n - 1) partial permutation matrix, and a matrix with exactly one off-main diagonal 1 entry.

DEFINITION 2.4.3. A Boolean matrix A is *prime* if whenever A = BC, B, C $\in B_n$ then B or C is a permutation, and A is not regular.

H. M. Devadze [89], D. J. Richman and H. Schneider [310], J. Borosh, D. J. Hartfiel and C. J. Maxon [33], D. de Caen and D. A. Gregory [46], and D. A. Gregory and N. J. Pullman [131] have independently established various properties of prime Boolean matrices. M. Tchuente [381] has constructed prime Boolean matrices from tree graphs.

DEFINITION 2.4.4. A matrix $A \in B_n$ is *fully indecomposable* if there do not exist $P, Q \in P_n$ such that PAQ has the block form

$$\begin{bmatrix} * & 0 \\ * & * \end{bmatrix}$$

It can be proved that a minimal generating set for B_n can be obtained by adding to the preceding four matrices one matrix from every D-class of prime matrices. A prime matrix is cogredient to a direct sum of an identity matrix and a fully indecomposable matrix. For more on fully indecomposable matrices, see Section 5.2.

THEOREM 2.4.1 [84]. There are at least $2^{n^2/4-O(n)}$ prime matrices in B_n.

Proof. Consider the matrices A such that $a_{ij} = 1$ if and only if (i) $(i, j) = (1, n)$ or (ii) $i = j$ or (iii) $n - 1 + j > i > j$ and $i - j$ is odd. These are fully indecomposable matrices because of the 1 entries a_{1n}, a_{ii}, $a_{i+1,i}$. The row and column ranks are n. No main diagonal 1 lies in a 2×2 rectangle of ones.

A factorization of a matrix A = BC expresses its ones as a union of n rectangles of ones. If each rectangle lay in a single row or column either B or C would be a permutation matrix or if both rows and columns occur, A would not be fully indecomposable. No two main diagonal entries lie in the same rectangle or they would lie in 2×2 rectangle of ones. So there is one main diagonal entry in each of the n rectangles. So some main diagonal 1 lies in a rectangle of size at least 2×2. This contradicts the fact that no main diagonal 1 lies in a 2×2 rectangle of ones. This completes the proof. □

A *quasi-variety* of binary relations (named by analogy with quasi-varieties of groups and semigroups) is the class of all binary relations for which a class of laws of the following forms holds:

(QV1) For all $(i, j) \in K$, $x_i \rho x_j$ implies $x_a \rho x_b$.

(QV2) For all $(i, j) \in K$, $x_i \rho x_j$ implies $x_a \rho^c x_b$.

(QV3) For all $(i, j) \in K$, $x_i \rho x_j$ implies $x_a = x_b$.

Such a law is specified by giving K, a, b, where K may be any subset of a Cartesian product of a set with itself. For instance if $K = \{(1, 2), (2, 3)\}$, $a = 1$, $b = 3$, (QV1) gives transitivity. In K. H. Kim and F. W. Roush [204] it is proved that the quasi-varieties are those classes of binary relations closed under direct products, restrictions to a subset, and isomorphism.

A *transformation* is a function from a set S to itself. A *partial transformation* is a function g from part of a set S to itself (or we can extend it to a transformation on $S \cup \{0\}$ by letting $g(x) = 0$ if g is not defined into S). The sets of transformations and partial transformations are semigroups under composition and can be regarded as subsemigroups of B_n. For instance if $S = \underline{n}$, the subsemigroup of B_n consisting of all matrices having no two ones in the same row is isomorphic to the semigroup of partial transformations. The rank of a partial transformation or transformation is the size of its image set. E. Howorka [159] proved that every infinite Boolean matrix is a product of a transposed transformation matrix and a transformation matrix (this is not true in the finite case). K. H. Kim and F. W. Roush [206] have investigated representations of arbitrary binary relations as sums of transformations and transposed transformations.

Transformations have been very extensively studied by semigroup theorists (see J. M. Howie [159] for instance). Every transformation can be represented by a directed graph in which each vertex has exactly one outgoing arrow. The restricted functions corresponding to the weak components of this graph are called the *generalized cycles* of the transformation. The *restriction* of a transformation f on S to $\bigcap_{n=1}^{\infty} f^n(S)$ is a permutation, called the *main permutation* of f. Two transformations are L (R, D)-equivalent if and only if they have the same image (induce the same equivalence relation on S by $f(x) = f(y)$, have the same rank). The equivalence relation f(x) = f(y) is called the *partition* associated with the transformation. It is known that the normal subsemigroups N of the semigroup of transformations are unions of {transformations of rank no more than m} and a normal subsemigroup of the symmetric group, where normal means if $x \in N$ then $pxp^{-1} \in N$ and $exe \in N$ for any permutation p and idempotent e. Every transformation has an inverse. Any semigroup is isomorphic to a subsemigroup of a semigroup of transformations. A semigroup is an *inverse semigroup* if and only if it is isomorphic to a subsemigroup of a semigroup of a semigroup of one-to-one partial transformations (partial permutations).

Many combinatorial results are known. See J. D. Dixon [92], P. Erdös and A. Renyi [104], P. Erdös and J. Spencer [105], M. Szalay [380], K. P. Kozlov [219, 220], Z. Todorov [385], J. Denes [70, 72-82], K. H. Kim and F. W. Roush [202, 206] and other works of these authors.

K. H. Kim and J. R. Krabill [181, 182] have studied maximal commutative subsemigroups of B_n. These include (i) the circulants, (ii) the diagonal matrices, (iii) matrices having more than half ones in each row and at least half ones in each column, and (iv) the transpose of (iii). K. H. Kim and S. Schwarz [210] have described all idempotents in the semigroup of circulants, and obtained counting results.

Two matrices A, B over a semiring are said to be *shift equivalent* if and only if there exist matrices R, S over the semiring such that for some nonnegative integer n, RA = BR, AS = SB, $SR = A^m$, $RS = B^m$. For nonsingular matrices over a field this concept deduces to similarity of A, B but over a semiring it is an important generalization of similarity.

Shift equivalence of matrices over the nonnegative integers has important applications to topological dynamics and the study of diffeomorphisms [6, 35-37, 259, 371, 401, 402]. By a study of shift equivalence of Boolean matrices we have shown that if A, B have no eigenvalues of multiplicity greater than one, there exists a finite procedure for deciding whether or not A, B are shift equivalent [207].

We will present a few results about shift equivalence of Boolean matrices in the exercises.

Exercises

1. Write out all the regular elements of B_2.
2. Write out all the idempotents of B_2.
3. Let A be a reduced element of B_n. Then show that $A \in \text{Reg}(B_n)$ if and only if $A = APA$ for some $P \in P_n$.
4. Compute $|\text{Idem}(B_3)|$.
5. Prove Lemma 2.1.5.
6. Prove Lemma 2.1.7.
7. Prove Lemma 2.1.8.
8. Prove Lemma 2.1.12.
9. Prove Proposition 2.1.14.
10. Prove Proposition 2.1.15.
11. Prove Lemma 2.1.18.
12. Prove Proposition 2.2.4.
13. Prove Proposition 2.2.5.
14. Find the largest inverse of the following matrices:

$$\begin{bmatrix} 1 & 0 & 1 \\ 0 & 1 & 0 \\ 0 & 1 & 1 \end{bmatrix}, \begin{bmatrix} 1 & 0 & 0 \\ 1 & 1 & 0 \\ 1 & 1 & 1 \end{bmatrix}, \begin{bmatrix} 0 & 0 & 0 \\ 0 & 0 & 1 \\ 1 & 0 & 0 \end{bmatrix}, \begin{bmatrix} 1 & 1 & 0 \\ 1 & 0 & 1 \\ 0 & 0 & 1 \end{bmatrix}$$

15. Prove Theorem 2.1.27.
16. Characterize those groups which are automorphism groups of $p \times p$ circulants where p is prime. Prove that the order of such a group is less than $p(p - 1)(p - 2)$ unless the circulant is 0, J, a permutation, or the complement of a permutation.
17. Suppose the circulant again is not 0, J, a permutation, or the complement of permutation, and suppose it has an automorphism of prime order $r \neq p$. Prove that r divides $p - 1$.
18. Prove that the class of maximal subgroups of semigroups of binary relations includes all finite groups.

19. Compute $\#_G(P_n)$.
20. Compute $\#_G(B_2)$.
21. How many maximal submonoids does B_3 have ?
22. Show that all maximal submonoids of B_n are simple semirings, i.e., there exist only two distinct congruences on them.
23. Write out all the prime matrices of B_2.
24. Compute the permanent of the matrices in Exercise 14.
25. Let S be the set of m-tuples of integers from 1 to k. Let x_i for i = 1 to k, be the mapping from S to S sending $(y_1, \ldots, y_m)$ to $(i, y_1, \ldots, y_{m-1})$. Prove distinct words in the x_i of length less than or equal to m are distinct transformations. Using this results, show that if R is a semiring containing 0 and at least one non-nilpotent element, that no nontrivial identity holds for all semigroups of square matrices over R.

In the following exercises, let T_n be the set of all tranformations defined on $\underline{n}$. Let $E_{n,m} = \{f \in T_n : \rho(f) \leq m\}$.

26. Compute the number of elements of the rank r D-class of T_n. (Use Stirling's numbers of the second kind). How many idempotents does this D-class have ?
27. Prove that $E_{n,m}$ is generated by the set of rank n idempotents.
28. Prove that all the normal subsemigroups of T_n are $E_{n,m}$.
29. Prove that all transformation matrices are regular, as follows. For a transformation f, and for all y contained in the image of f, choose an element g(y) such that f(g(y)) = y. For all element z not in the image of f, let g(z) be one of the elements of g(y). Prove that fgf = f, gfg = g. Thus row rank, column rank, Schein rank coincide for transformations, and the rank of a product cannot exceed the rank of the factors.
30. Prove that the rank of a transformation matrix equals the cardinality of the image of the tranformation.
31. Prove that if X is infinite every binary relation on X is the composition of a transformation with a transposed transformation.

In what follows let P_n^*, T_n^* denote the semigroups of partial permutations, and partial transformations on $\underline{n}$.

32. Describe the R, L, D, H, J-classes in P_n, T_n, P_n^*, T_n^*.

33. Prove $|P_n| = n!$, $|T_n| = n^n$, $|P_n^*| = \sum_{i=0}^{n} i!\binom{n}{i}^2$, and $|T_n^*| = \sum_{i=0}^{n} \binom{n}{i} n^i$.

34. Prove all the maximal subgroups of T_n are symmetric groups.

35. Prove T_3 has subsemigroups of all orders from 1 to 27 except 18, 19, 20, 25, 26.

36. Show that the circulants form a commutative subsemigroup of B_n. This is denoted C_n.

37. Prove that C_n is a maximal commutative subsemigroup of B_n.

38. Prove that in C_n, A L B if and only if $P^m A = B$ for some m. Why are the relations L, R the same in C_n ?

39. Let $D_n = \{A \in B_n :$ for all $i \neq j$, $a_{ij} = 0\}$. Prove D_n is a maximal commutative subsemigroup of B_n.

40. What is the center of B_n, i.e., which Boolean matrices commute with all others ?

Let F_n^r denote the collection of all $n \times n$ Boolean matrices in which each row has at least half ones and each column has more than half ones.

41. Prove the product of any two matrices of F_n^r is J.

42. Prove that F_n^r together with 0, I forms a maximal commutative subsemigroup of B_n.

43. Define a subsemigroup analogous to F_n^r in which the roles of rows and columns are interchanged.

44. Find all maximal commutative subgroups of B_2, and if possible of B_3.

45. For $n = 2^k$ list the nonzero idempotents of C_n.

46. Describe shift equivalence in the semigroup of T_n.

47. Describe shift equivalence for any two Boolean matrices.

48. Show that the following two Boolean matrices are shift equivalent:

$$\begin{bmatrix} 1 & 0 & 0 \\ 0 & 0 & 0 \\ 1 & 0 & 1 \end{bmatrix}, \begin{bmatrix} 1 & 0 & 0 \\ 1 & 1 & 1 \\ 1 & 1 & 1 \end{bmatrix}$$

49. Prove that two idempotent Boolean matrices are shift equivalent if and only if they belong in the same D-class.

CHAPTER THREE

INVERSES

In this chapter we investigate various inverses of Boolean matrices. Only since 1971 has this field of study of inverses been systematically studied and explored, though the concept of inverse of Boolean matrices was first introduced by J. H. M. Wedderburn [397] as early as 1934 as follows.

A square Boolean matrix B is said to be an *inverse* of a square Boolean matrix A if $AB = BA = I$. In 1952, R. D. Luce [234] showed that A possesses a two-sided inverse if and only if A is an orthogonal matrix in the sense that $AA^T = I$, and that, in this case, A^T is a two-sided inverse. In 1963, D. E. Rutherford [330] showed that if A possesses a one-sided inverse, that inverse is also a two-sided inverse. Furthermore such an inverse, if it exists, is unique and is A^T. It is easy to note that the permutation matrices are the only matrices which have such inverses.

Over the past two decades many interesting and important results on the various inverses of matrices over the complex field have been developed. However, till recently, not much work has been done in this regard for matrices over algebras which are not fields. R. J. Plemmons [282] and C. R. Rao [294] explored this new field of study and suggested some applications to graph theory and network analysis.

3.1 Generalized Inverse

The notion of generalized inverse of an arbitrary matrix apparently originated in the work of E. H. Moore [258]. R. D. Sheffield [362] following K. O. Friedrich [116] using a similar idea and a weaker definition uses the name *pseudoinverses* to describe a linear operator having certain properties possessed by an inverse. More recently, R. Penrose [277], C. A. Desoer and B. H. Whalen [86], A. Ben-Israel and A. Charnes [18], Y. Ijiri [160], and J. E. Scroggs and P. L. Odell [354] have further explored the generalized inverse from the various points of view. Although the definitions, which are equivalent, given by Penrose and others insure the uniqueness of the generalized inverse, an important spectral property of the inverse of a nonsingular matrix is not inherited by the generalized inverse of a singular matrix. The reader can find full account of this and related work in A. Ben-Israel and T. N. E. Greville [19], M. Z. Nashed [261], C. R. Rao and S. K. Mitra [295].

DEFINITION 3.1.1. Let $A \in B_{mn}$. Then $G \in B_{nm}$ is said to be a *generalized inverse* (*g-inverse*) of A, denoted by A^- if $A = AGA$. A g-inverse G of A is said to be a *maximum g-inverse* (*greatest g-inverse*) of A if every $A^- \leq G$.

Note that a g-inverse of a matrix is not necessarily unique. And that since a sum of g-inverses is a g-inverse, there will be a maximum g-inverse whenever at least one g-inverse exists. We will usually assume $A \neq 0$.

Here we present three approaches to g-inverses, due respectively to P. P. Rao [296], K. H. Kim and F. W. Roush [197], and R. J. Plemmons [282].

Rao's method is based on the following observation. Suppose we rearrange the rows and columns of any matrix so that the first r rows and the first s columns form a row basis and a column basis respectively. Then the matrix has the block form:

$$\begin{bmatrix} A_1 & A_1C \\ DA_1 & DA_1C \end{bmatrix}$$

where A_1 is the submatrix formed by the first r rows and first s columns. This is true for any matrix over any semiring. Also all rows and all columns of A_1 will be independent. If A is regular, A_1 is a square matrix.

Rao next shows that any g-inverse of A_1, yields a g-inverse of A. He shows that if A_1 has a g-inverse, then some permutation matrix is a g-inverse of A_1. Thus any regular matrix has a g-inverse which is a partial permutation matrix. Symmetric idempotent matrices can readily be classified. They are the matrices of partial equivalence relations and can put into the form of a direct sum of J matrices and 0 matrices of various dimensions.

Rao then proves that A_1 has a g-inverse if and only if its permanent is 1 and $|\{j: a_{ij} = 1\}| = |\{j: A_{j*} \le A_{i*}\}|$. Recall from Section 2.1 that rank n idempotents satisfy these conditions. Essentially A_1 is a rank n idempotent with its rows rearranged. From this he derives an algorithm which yields a g-inverse if one exists.

In the process we view regularity and idempotence from a different angle than in Section 2.1.

We now present necessary and sufficient conditions of existence of g-inverses of rectangular matrices.

DEFINITION 3.1.2. A matrix $A \in B_{mn}$ is called a *Rao decomposable matrix* if there exist permutation matrices P and Q and matrices C and D such that

$$PAQ = \begin{bmatrix} A_1 & A_1C \\ DA_1 & DA_1C \end{bmatrix}$$

(This concept was first introduced by P. P. Rao [296]).

For brevity, we let

$$[A_1\text{-}C\text{-}D] = \begin{bmatrix} A_1 & A_1C \\ & \\ DA_1 & DA_1C \end{bmatrix}$$

THEOREM 3.1.1 [296]. Let $A \in B_{mn}$ with $\rho_r(A) = r$ and $\rho_c(A) = c$. Then A is a Rao decomposable matrix with $A_1 \in B_{rc}$, $\rho_r(A_1) = r$, and $\rho_c(A_1) = c$.

Proof. Let Q be a permutation matrix such that the first c columns of AQ form the basis of C(A), i.e., $AQ = [B \quad BC]$ for some C, where $B \in B_{mc}$ with $\rho_c(B) = c$.

By Corollary 2.1.11, $\rho_r(B) = \rho_r(A) = r$. Thus there exists a permutation matrix P such that first r rows of PB form the basis of $C(B^T)$, i.e., $PB = [A_1 \quad DA_1]^T$ for some D where $A_1 \in B_{rc}$ with $\rho_r(A_1) = r$. It follows that $\rho_c(A_1) = \rho_c(B) = c$. Thus $PAQ = P[B \quad BC] = [PB \quad PBC] = [A_1\text{-}C\text{-}D]$ for some C and D where $A_1 \in B_{rc}$ with $\rho_r(A_1) = r$ and $\rho_c(A_1) = c$. This proves the theorem. □

We remark that in the above decomposition $B_c(A) = P^T[A_1 \quad DA_1]^T$ and $B_r(A) = [A_1 \quad A_1C]Q^T$.

DEFINITION 3.1.3. A matrix $A \in B_{mn}$ is said to have *full rank* if $\rho_r(A) = m$ and $\rho_c(A) = n$.

THEOREM 3.1.2 [296]. Let A be a Rao decomposable Boolean matrix and A_1 be a full rank matrix. Then the following statements are equivalent. (i) A^- exists. (ii) $([A_1 \quad DA]^T)^-$ exists. (iii) $[A_1 \quad A_1C]^-$ exists. (iv) A_1^- exists.

Proof. Given a g-inverse of any of the above four matrices, instead of just showing the existence of g-inverses of the rest we will construct g-inverses for the rest. Proofs are by straightforward verifications.

(i) implies (ii)-(iv). If A^- exists, then $(PAQ)^-$ exists, say

$$(PAQ)^- = \begin{bmatrix} G_1 & G_2 \\ G_3 & G_4 \end{bmatrix}$$

This implies $[G_1 + CG_3 \quad G_2 + CG_4] = [A_1 \quad DA_1]^-$, $[G_1 + G_2D \quad G_3 + G_4D]^T = [A_1 \quad A_1C]^-$, and $[G_1 + CG_3 + G_2D + CG_4D] = A_1^-$.

(ii) implies (iii), (iv), and (i). Let $[G_1 \quad G_2] = ([A_1 \quad DA_1]^T)^-$, then $[G_1 + G_2D \quad 0]^T = [A_1 \quad A_1C]^-$, $[G_1 + G_2D] = A_1^-$, and

$$A^- = Q\begin{bmatrix} G_1 & G_2 \\ 0 & 0 \end{bmatrix}P$$

(iii) implies (iv), (i), and (ii). Let $[G_1 \quad G_2]^T = [A_1 \quad A_1C]^-$, then $[G_1 + G_2D] = A_1^-$,

$$A^- = Q\begin{bmatrix} G_1 & 0 \\ G_2 & 0 \end{bmatrix}P$$

and $[G_1 + CG_2 \quad 0] = ([A_1 \quad DA_1]^T)^-$.

(iv) implies (i)-(iii). Let $G_1 = A_1^-$, then

$$A^- = Q\begin{bmatrix} G_1 & 0 \\ 0 & 0 \end{bmatrix}P$$

$[G_1 \quad 0] = ([A_1 \quad DA_1]^T)^-$ and $[G_1 \quad 0]^T = [A_1 \quad A_1C]^-$. This proves the theorem. □

COROLLARY 3.1.3. Let $A, B \in B_{mn}$. If $C(A) = C(B)$, then A^- exists if and only if B^- exists.

Note that in the proof of Theorem 3.1.2, $[A_1\text{-}C\text{-}D]^- = (PAQ)^-$ if and only if $[G_1 + CG_3 + G_2D + CG_4D] = A_1^-$. Thus the existence of a g-inverse of a matrix reduces to the problem of existence of g-inverses of full rank matrices.

PROPOSITION 3.1.4. If $A \in B_{mn}$ has a g-inverse, then (i) $\rho_c(A) = n$ implies $n \leq m$, (ii) $\rho_r(A) = m$ implies $m \leq n$.

Proof. Let $G = A^-$. Then $C(A) = C(AG)$. This implies that all the independent columns, i.e., all the columns of A are available among the columns of AG. But the order of AG is $m \times m$ and so $n \leq m$. A similar proof holds for the row rank. □

REMARK 3.1.1. If A is a full rank Boolean matrix and it has a g-inverse then A is square.

COROLLARY 3.1.5. If $A \in B_{mn}$ has a g-inverse, then $\rho_c(A) = \rho_r(A)$.

Proof. Let $\rho_c(A) = c$ and $\rho_r(A) = r$. Then by Theorem 3.1.2, A is a Rao decomposable matrix where A_1 is an $r \times c$ full rank matrix. Again by Theorem 3.1.2, A^- exists and so A_1^- exists. This implies A_1 is square by Remark 3.1.1. Thus $c = r$, i.e., $\rho_c(A) = \rho_c(A)$. □

PROPOSITION 3.1.6. If $A \in B_{mn}$ has a g-inverse, then (i) $\rho_r(A) = m$ implies $AG_1 = AG_2$ for all g-inverses G_1 and G_2, (ii) $\rho_c(A) = n$ implies $G_1A = G_2A$ for all g-inverses G_1 and G_2.

REMARK 3.1.2. In case of real matrices, if the class of all g-inverses of A and B are same then A = B. But note that in case of Boolean matrices this result is not even true if A^- and B^- exist. For example, consider

$$\begin{bmatrix} 1 & 0 & 0 & 0 & 1 & 1 \\ 0 & 1 & 0 & 1 & 0 & 1 \\ 0 & 0 & 1 & 1 & 1 & 0 \\ 0 & 1 & 1 & 1 & 1 & 1 \\ 1 & 0 & 1 & 1 & 1 & 1 \\ 1 & 1 & 0 & 1 & 1 & 1 \end{bmatrix}, \begin{bmatrix} 1 & 0 & 0 & 1 & 1 & 1 \\ 0 & 1 & 0 & 1 & 1 & 1 \\ 0 & 0 & 1 & 1 & 1 & 1 \\ 1 & 1 & 1 & 1 & 1 & 1 \\ 1 & 1 & 1 & 1 & 1 & 1 \\ 1 & 1 & 1 & 1 & 1 & 1 \end{bmatrix} \in B_6$$

THEOREM 3.1.7 [296]. Let A be a full rank Boolean matrix such that A^- exists. Then there exists a unique permutation matrix P which is a g-inverse of A.

Proof. Since A is a full rank Boolean matrix and A^- exists, A is square by Remark 3.1.1. Let $G = A^-$. Then $C(A) = C(AG)$ and AG is idempotent. Since A is a full rank matrix, the columns of AG are nothing but a permutation of the columns of A, i.e., there exists a permutation matrix P such that $AP = AG$ and so $APA = AGA = A$ which in turn implies that $P = A^-$.

To show uniqueness of P, if possible let P_1 and P_2 be two permutation matrices which are g-inverses of A. Then by Proposition 3.1.6, $AP_1 = AP_2$, and since columns of A are distinct because A is full rank, $P_1 = P_2$. This proves the theorem. □

COROLLARY 3.1.8. If $A \in B_{mn}$ has a g-inverse, then A is a Rao decomposable matrix such that A_1 is a full rank matrix and A_1 is an idempotent matrix.

The following result is a generalization of Theorem 3.1.7 to any matrix which has a g-inverse.

THEOREM 3.1.9 [296]. Let $A \in B_{mn}$ has a g-inverse, then there exists a partial permutation matrix which is a g-inverse.

Proof. By Corollary 3.1.8, there exist permutation matrices P_1 and Q_1 such that $P_1AQ_1 = [A_1\text{-}C\text{-}D]$ where A_1 is idempotent. Then the partial transformation matrix

$$P = Q_1 \begin{bmatrix} I & 0 \\ 0 & 0 \end{bmatrix} P_1$$

is a g-inverse of A. This proves the theorem. □

We note that the partial permutation matrix P in the above theorem is not unique. (Let A = J).

Before proceeding to the next section a couple of results on idempotent matrices are stated below.

PROPOSITION 3.1.10. A matrix $A \times B_n$ is an idempotent matrix of rank r if and only if there exists a permutation matrix P such that $PAP^T = [A_1\text{-}C\text{-}D]$ where A_1 is an $r \times r$ full rank idempotent matrix and C and D are such that $CD \leq A_1$.

Proof. Necessity. Let A be a rank r idempotent matrix. From Lemma 2.1.6 it follows that if we delete dependent rows and columns we get an idempotent matrix B_1. Let P be the permutation matrix which makes the remaining rows the first ones. Existence of C, D follows from the dependence of the other rows and columns on the first ones. Proof of the remaining conditions is a computation.

Sufficiency. This is also a computation. □

COROLLARY 3.1.11. If A is a symmetric and full rank idempotent Boolean matrix, then A = I.

COROLLARY 3.1.12. A Boolean matrix A is a symmetric idempotent matrix of rank r if and only if there exists a permutation matrix P such that

$$\begin{bmatrix} B_1 & 0 & 0 & \dots & 0 \\ 0 & B_2 & 0 & \dots & 0 \\ & & \dots\dots\dots & & \\ 0 & & 0\ B_r & \dots & 0 \\ & & \dots\dots\dots & & \\ 0 & 0 & 0 & \dots & 0 \end{bmatrix}$$

in block form. Here the submatrix B_i is the matrix of a reflexive, symmetric, and transitive relation, i.e., an equivalence relation.

THEOREM 3.1.13 [296]. Let $A \in B_{mn}$ be a reflexive matrix. If $G = A^-$, then $g_{ij} \leq a_{ij}$ for all $i, j \in \underline{t} \subset \underline{n}$, where $|\underline{t}| = \min\{m, n\}$.

Proof. Let $a_{ij} = 0$ for some $i, j \leq s \leq m$, $s \leq n$, and $A = AGA$. This implies

$$a_{ij} = \sum_{k=1}^{n} \sum_{h=1}^{m} a_{ik} g_{kh} a_{hj} = 0$$

Thus $a_{ik} g_{kh} a_{hj} = 0$ for all $k \in \underline{n}$, $h \in \underline{m}$ so $a_{ii} g_{ij} a_{jj} = 0$ which in turn implies that $g_{ij} = 0$. Thus $g_{ij} \leq a_{ij}$ for all $i, j \leq s$. □

Note that in particular, if A is a reflexive and idempotent Boolean matrix, then the maximum g-inverse of A is itself. Moreover if A is a full rank idempotent Boolean matrix, then $G = A^-$ if and only if $I \leq G \leq A$.

The following is a generalization of Theorem 2.1.16 to rectangular Boolean matrices.

DEFINITION 3.1.4. A matrix $A \in B_{mn}$ is said to be *space decomposable* if for some k there exist two matrices X and Y of orders $m \times k$ and $k \times n$ respectively such that (i) $A = XY$, (ii) $C(A) = C(X)$, and (iii) $R(A) = R(Y)$. This decomposition will be called a *space decomposition* of A.

Note that if $A \in \text{Reg}(B_n)$, then A is space decomposable.

THEOREM 3.1.14 [296]. Let $A \in B_{mn}$. Then A^- exists if and only if A is space decomposable.

Proof. Necessity. Assume A^- exists. Then A is of the form $P\ A_1\text{-}C\text{-}D\ Q$ where A_1 is a full rank idempotent matrix, $\rho(A) = r$, and P, Q are permutation matrices. Thus $A = P[A_1 \quad DA_1]^T[A_1 \quad A_1C]Q = XY$, where $X = P[A_1 \quad DA_1]^T$, $Y = [A_1 \quad A_1C]Q$. It is easy to see that $C(A) = C(X)$ and $R(A) = R(Y)$, and so $A = XY$ is a space decomposition.

The converse follows from Theorem 2.1.17. □

The next four propositions are due to P. P. Rao [296].

PROPOSITION 3.1.15. Let $A = P[A_1\text{-}C\text{-}D]Q$ where P and Q are permutation matrices and A_1 is a full rank idempotent matrix. Then $A = X_1Y_1$ is a space decomposition of A if and only if $X_1 = XP_1$ and $Y_1 = P_1^TY$ for some permutation matrix P where $X = P[A_1 \quad DA]^T$ and $Y = [A_1 \quad A_1C]Q$.

Proof. Exercise.

PROPOSITION 3.1.16. Let A = XY be a space decomposition of $A = P[A_1\text{-}C\text{-}D]Q$ and $G = A^-$ and A_1 be a nonsingular idempotent matrix. Then (i) X^-, Y^- exist, (ii) $X^-X = YY^-$, (iii) $X^-A = Y$ and $AY^- = X$, (iv) $Y^-X^- = A^-$, (v) $YA^- = X^-$ and $A^-X = Y^-$ for $A^- = Y^-X^-$.

DEFINITION 3.1.5. A set of vectors $S_v = \{v_1, \ldots, v_n\}$ is said to satisfy the *weight condition* or *condition* w_c, if for every i, there are exactly w_i vectors in the set which are less than or equal to v_i where w_i is the weight of the vector v_i.

PROPOSITION 3.1.17. Let A be a full rank square Boolean matrix. Then A^- exists if and only if Per (A) = 1 and the columns of A satisfy weight condition.

Proof. Exercise.

COROLLARY 3.1.18. Let A be as in Proposition 3.1.17. Then A^- exists if and only if Per (A) = 1 and the rows of A satisfy weight condition.

COROLLARY 3.1.19. Let A be as in Proposition 3.1.17 and assume A^- exists. Then there exists a column of A of weight one.

COROLLARY 3.1.20. Let A be a full rank square Boolean matrix such that Per (A) = 1 and such that the columns of A satisfy condition w_c. Let $i_1, \ldots, i_n$ be such that $a_{1i_1} a_{2i_2} \cdots a_{ni_n} = 1$ then

$$A^- = [e^{i_1} \; e^{i_2} \; \cdots \; e^{i_n}]$$

PROPOSITION 3.1.21. Let A be a full rank square Boolean matrix such that

$$PAQ = \begin{bmatrix} 1 & v^T \\ 0 & A_1 \end{bmatrix}$$

where P and Q are permutation matrices and v and 0 are column vectors. If A^- exists then A_1^- exists and A_1 is a full rank matrix.

Proof. Exercise.

COROLLARY 3.1.22. If $A \in B_{mn}$ is of full rank and A^- exists then there exist permutation matrices P and Q such that PAQ is of full rank, idempotent, and upper triangular matrix.

We now present an algorithm due to P. P. Rao [296]. Let $A \in B_{mn}$ and suppose A^- exists. For computing A^- we proceed as follows. First we obtain the row and the column basis of A. Then we can compute the permutation g-inverses of the submatrix with full rank formed by these rows and columns of A. Finally A^- is constructed as in Theorem 3.1.2.

ALGORITHM 3.1.1. Let $A \in B_{mn}$.
Step 1. Let w_i be the weight of A_{i*}.
Step 2. Let B = A.

Step 3. Perform the following construction on B. Choose a nonzero row of minimum weight in B. If there is more than one row in B of this weight, choose one of these rows B_{r*} such that w_r is a minimum. Delete all columns B_{*j} such that $b_{rj} = 1$. Go back to Step 3. (This step, for a regular matrix A, simply finds a row basis for A).

Step 4. Let C be the matrix such that $C_{r*} = A_{r*}$ if A_{r*} was some row considered in Step 3, and $C_{r*} = 0$ otherwise. (In other words, delete all dependent rows from A).

Step 5. Let D = C. Let P = 0.

Step 6. Choose a nonzero column in D of minimum weight. If there is more than one column of D having minimum weight choose one of these which has minimum weight in C. Let this column be D_{*j}. Then for some k such that $d_{kj} = 1$, let $p_{kj} = 1$. Delete all rows D_{m*} such that $d_{mj} = 1$. If D = 0 go to Step 6. Else go back to Step 5. (This step finds a partially ordered column basis for D).

Step 7. If $AP^TA = A$ then P^T is a g-inverse of A. Otherwise A is not regular. (Actually P is composed of identification vectors. To see that this step is correct, assume A is regular and we have deleted all dependent rows and columns of A and rearranged the rows and columns so that A is a nonsingular, reflexive idempotent. Then P will be the identity matrix which is in fact a g-inverse).

We remark that Rao actually works with the single matrix D + P in Step 5. On a computer this would save some computer space for very large matrices.

EXAMPLE 3.1.1. Let

$$A = \begin{bmatrix} 1 & 0 & 1 & 1 \\ 1 & 0 & 1 & 0 \\ 0 & 0 & 0 & 1 \\ 1 & 1 & 1 & 1 \\ 0 & 1 & 0 & 0 \end{bmatrix}$$

Taking r = 3, we get

$$B = \begin{bmatrix} 1 & 0 & 1 & 0 \\ 1 & 0 & 1 & 0 \\ 0 & 0 & 0 & 0 \\ 1 & 1 & 1 & 0 \\ 0 & 1 & 0 & 0 \end{bmatrix}$$

Taking r = 5, we get

$$B = \begin{bmatrix} 1 & 0 & 1 & 0 \\ 1 & 0 & 1 & 0 \\ 0 & 0 & 0 & 0 \\ 1 & 0 & 1 & 0 \\ 0 & 0 & 0 & 0 \end{bmatrix}$$

Here all three non-null rows of B are of equal weight. But second row has minimum original weight. So we take r = 2. Then B becomes null. We form now

$$D = \begin{bmatrix} 0 & 0 & 0 & 0 \\ 1 & 0 & 1 & 0 \\ 0 & 0 & 0 & 1 \\ 0 & 0 & 0 & 0 \\ 0 & 1 & 0 & 0 \end{bmatrix}$$

Now taking j = 1, we get

$$D = \begin{bmatrix} 0 & 0 & 0 & 0 \\ 0 & 0 & 0 & 0 \\ 0 & 0 & 0 & 1 \\ 0 & 0 & 0 & 0 \\ 0 & 1 & 0 & 0 \end{bmatrix} \qquad P = \begin{bmatrix} 0 & 0 & 0 & 0 \\ 1 & 0 & 0 & 0 \\ 0 & 0 & 0 & 0 \\ 0 & 0 & 0 & 0 \\ 0 & 0 & 0 & 0 \end{bmatrix}$$

If j = 2, then

$$D = \begin{bmatrix} 0 & 0 & 0 & 0 \\ 0 & 0 & 0 & 0 \\ 0 & 0 & 0 & 1 \\ 0 & 0 & 0 & 0 \\ 0 & 0 & 0 & 0 \end{bmatrix} \qquad P = \begin{bmatrix} 0 & 0 & 0 & 0 \\ 1 & 0 & 0 & 0 \\ 0 & 0 & 0 & 0 \\ 0 & 0 & 0 & 0 \\ 0 & 1 & 0 & 0 \end{bmatrix}$$

If $j = 4$, then

$$D = 0, \qquad P = \begin{bmatrix} 0 & 0 & 0 & 0 \\ 1 & 0 & 0 & 0 \\ 0 & 0 & 0 & 1 \\ 0 & 0 & 0 & 0 \\ 0 & 1 & 0 & 0 \end{bmatrix}$$

Now it can easily be checked that $P^T = A^-$.

Next we present an algorithm due to K. H. Kim and F. W. Roush [197]. This yields all g-inverses, and in fact, a formula for the number of g-inverses. For hand computation we think it is a little easier than Rao's method, though his would work well on a computer.

It is based on the concept of identification vectors (see Definition 2.1.13). Let A_{i*} be a row basis vector of a matrix A. Then an identification vector for A_{i*} is a vector u with only one 1, such that $u \le A_{i*}$ and for any j, $u \le A_{j*}$ if and only if $A_{j*} \ge A_{i*}$. In Lemma 2.1.28 we showed that if a Boolean matrix is regular then every row basis vector has an identification vector. Here we establish the converse.

The relationship of this with Rao's method is that the set of identification vectors, transposed, gives a partial permutation matrix which is a g-inverse of the original matrix.

DEFINITION 3.1.6. Let $p(t) = \inf \{w \in R(A): w \ge t\}$ where the infimum is taken in R(A).

ALGORITHM 3.1.2. Let $A \in B_n$.

Step 1. Find $B_r(A)$.

Step 2. Find I(v) for each $v \in B_r(A)$. A g-inverse will exist if and only if I(v) is nonempty for each $v \in B_r(A)$. (Recall Definition 2.1.13 that I(v) is the set of identification vectors of $B_r(A)$).

Step 3. For each $v \in B_r(A)$, choose an identification vector $u \in I(v)$.

Step 4. For each such chosen u, choose a vector s (to be the image of u) such that $s_i = 1$ only if $A_{i*} \leq v$ and such that $s_i = 1$ for at least one i such that $A_{i*} = v$. (We sometimes denote s as s(v)).

Step 5. For any vector t with exactly one 1 entry other than the u's chosen in Step 3, if t is not less than any row vector, send t to an arbitrary vector. Otherwise send t to a vector q such that $q_i = 1$ only if $A_{i*} \leq p(t)$.

Step 6. Linearly order the set of vectors with only one 1 in the order in which they occur as row vectors of the identity matrix. Write the vectors s and q in the order of the vectors u and t.

EXAMPLE 3.1.2. Let

$$A = \begin{bmatrix} 1 & 1 & 1 \\ 0 & 1 & 1 \\ 1 & 0 & 0 \end{bmatrix}$$

Step 1. $B_r(A) = \{(0\ 1\ 1),\ (1\ 0\ 0)\}$.

Step 2. $I(0\ 1\ 1) = \{(0\ 1\ 0),\ (0\ 0\ 1)\}$, $I(1\ 0\ 0) = \{(1\ 0\ 0)\}$.

Step 3. There are two ways to choose the u's.

Step 4. For u = (1 0 0), s = (0 0 1). For the other u chosen, s = (0 1 0).

Step 5. For the remaining 1 element vector t, which will be one of (0 1 0), (0 0 1), p(t) = (0 1 1). The vector q can be either 0 or (0 1 0).

Step 6. There are 3 g-inverses of A:

$$\begin{bmatrix} 0 & 0 & 1 \\ 0 & 1 & 0 \\ 0 & 0 & 0 \end{bmatrix}, \quad \begin{bmatrix} 0 & 0 & 1 \\ 0 & 1 & 0 \\ 0 & 1 & 0 \end{bmatrix}, \quad \begin{bmatrix} 0 & 0 & 1 \\ 0 & 0 & 0 \\ 0 & 1 & 0 \end{bmatrix}$$

The proof the next theorem is completely straightforward, using inqualities and the definition of identification vector. Note that ABA = A if and only if for each vector $v \in B_r(A)$, vBA = v.

THEOREM 3.1.24 [197]. Let $A \in B_n$. Then Algorithm 3.1.2 for all allowable choices yields all g-inverses of A.

Proof. First we will show matrices G constructed in this way are always g-inverses of A. Let $v \in B_r(A)$. Then vG is the sum of s(v), of $s(v_h)$ for all other basis vectors v_h whose identification vectors are less than or equal to v and of q(t) for all $t \le v$. By the definition of s, $s(v)A = v$ and $s(v_h)A = v_h$. By the definition of an identification vector, $v_h < v$ for each v_h. And by definition of q, $q(t)A \le p(t)$ and $p(t) \le v$ for each $t \le v$. Thus $vGA = v + \sum_{t \le v} p(t) + \sum v_h = v$. Thus for any sum v_1 of basis vectors in R(A), $v_1GA = v_1$. Thus $AGA = A$.

Now let G be any g-inverse of A. Let t be a vector less than or equal to a basis vector v. We have $tG \le vG$. But since $v = v_2A$ for some vector v_2, $(vG)A = v_2AGA = v_2A = A$. Thus $(tG)A \le v$, and so $(tG)A \le \inf_{t \le w} w = p(t)$. This in turn implies $(tG)_i = 1$ only if $A_{i*} \le p(t)$.

This implies also for any identification vector u, that $(uG) \le v$ and $(uG)_i = 1$ only if $A_{i*} \le v$. Suppose that for all identification vectors $u \in I(v)$, $(uG)_i = 1$ only if $A_{i*} < v$. Then (uG)A is a sum of vectors properly less than v.

Let t be a vector with only one 1 which is less than or equal to v but not an identification vector for v. Then there exists $w \in B_r(A)$ such that $t \le w$ and $v \not\le w$. Else t would be an identification vector for v. Thus p(t) will be properly less than v. Thus (tG)A will be properly less than v. Thus $(vG)A = \sum_{u \le v} (vG)A + \sum_{t \le v} (tG)A$ will be a sum of vectors properly less than v. But $(vG)A = v$. This contradicts the assumption that v is a basis vector. Thus for some identification vector u of v, there will be some i such that $(uG)_i = 1$ and $A_{i*} = v$. This completes the proof. □

COROLLARY 3.1.25. Let $A \in B_n$. That $I(v) \ne 0$ for each $v \in B_r(A)$ is a necessary and sufficient condition for A to be regular.

COROLLARY 3.1.26. The number of g-inverses of A is equal to

$$\prod_{v \in B_r(A)} \left(2^{n(v)c(v)}\right)\left(2^{m(v)c(v)} - 1\right) \prod_{t \in T} 2^{n(t)}$$

where $m(v)$ is the number of rows of A equal to v, $n(v)$ is the number of rows of A less than v, $c(v)$ is the number of identification vectors v, T is the set of all vectors with exactly one 1 which are not identification vectors of any basis vector, and $n(t)$ is the number of rows of A which are less than or equal to $p(t)$ or $n(t) = n$ if $p(t) = \inf \{\emptyset\}$. Here $p(t)$ is the same as in Definition 3.1.6.

R. J. Plemmons [282] uses a more conceptual and less computational method of studying g-inverses.

PROPOSITION 3.1.27 [282]. Let $A \in B_n$. If A is a full rank matrix, then A has a g-inverse if and only if $A = APA$ for some $P \in P_n$. (Note that this proposition is basically the same as Theorem 3.1.7).

Proof. Suppose A is full rank and has a g-inverse. Then L_A contains an idempotent by Proposition 2.1.14 which will be a full rank idempotent E. Now since A and E are row independent in L_A, $E = PA$ for some $P \in P_n$. But $AE = A$, so $A = APA$.

The converse is obvious and a dual proof holds when A is column independent. This completes the proof. □

We next consider the existence of a g-inverse in terms of conditions on the rows and columns.

COROLLARY 3.1.28. If $A \in B_n$ has the property that for $i, j \in \underline{n}$, $A_{i*} + A_{j*} = A_{j*}$ $(A_{*i} + A_{*j} = A_{*j})$ implies $i = j$, then A is either a permutation matrix or else A has no g-inverse for $n > 1$.

3.2 Minimum Norm g-Inverses and Least Squares g-Inverses

A minimum norm (least squares) g-inverse is a g-inverse such that GA (AG) is a symmetric idempotent. This added symmetry condition restricts the class of matrices having such inverses. In fact if a matrix has either of these inverses, it is D-equivalent to a diagonal matrix. However this is not sufficient: any two row basis vectors must have product 0.

We give conditions of Rao for existence of these inverses and then our own conditions and algorithms for finding all minimum norm g-inverses and least squares g-inverses. We compute the number of matrices having a minimum norm g-inverses.

DEFINITION 3.2.1. Let $A \in B_{mn}$. Then $G \in B_{nm}$ is said to be a *minimum norm g-inverse* of A, denoted by A_m^- if $A = AGA$, $GA = (GA)^T$. Now G is said to be a *least squares g-inverse* of A, denoted by A_l^- if $A = AGA$, $AG = (AG)^T$.

THEOREM 3.2.1 [296]. Let $A \in B_{mn}$. If $A \neq 0$, then the following statements are equivalent. (i) A_m^- exists. (ii) A is of the form

$$P\begin{bmatrix} I & C \\ D & DC \end{bmatrix}Q$$

where P and Q are permutation matrices and C is such that $CC^T \leq I$. (iii) A^- exists and $C(A) = C(AA^T)$. (iv) There exists a G such that $GAA^T = A^T$.

Proof. (i) implies (ii). Let $G = A_m^-$. Then GA is symmetric and idempotent. Thus there exists a permutation matrix Q such that Corollary 3.1.12

$$GA = Q\begin{bmatrix} I & C \\ C^T & C^TC \end{bmatrix}Q^T$$

where C is such that $CC^T \leq I$. Again $G = A^-$ implies $C(A^T) = C((GA)^T)$

which implies A is of the form

$$P\begin{bmatrix} I & C \\ D & DC \end{bmatrix}Q$$

for some D.

(ii) implies (iii). It can be easily checked that

$$A_m^- = Q^T\begin{bmatrix} I & 0 \\ C^T & 0 \end{bmatrix}P^T$$

and $C(A) = C(AA^T)$.

(iii) implies (iv). If A^- exists and $C(A) = C(AA^T)$, then $(AA^T)^-$ exists and there exists a matrix D such that $A = AA^TD$. Now it is easy to check that $GAA^T = A^T$ where $G = D^T$.

(iv) implies (i). If $CAA^T = A^T$ then $GA = GAA^TG^T = A^TG^T$. Thus $A^T = GAA^T = A^TG^TA^T$. These implies $G = A_m^-$. This proves the theorem. □

THEOREM 3.2.2 [296]. Let $A \in B_{mn}$. If $A \neq 0$, then the following statements are equivalent. (i) A_l^- exists. (ii) A is of the form

$$P\begin{bmatrix} I & C \\ D & DC \end{bmatrix}Q$$

where P and Q are permutation matrices and D is such that $D^TD \leq I$. (iii) A^- exists and $C(A^T) = C(A^TA)$. (iv) There exists a G such that $A^TAG = A^T$.

Proof. Similar to the proof of Theorem 3.2.1. □

We present algorithms for finding all minimum norm g-inverses and least squares g-inverses. The following results are due to K. H. Kim and F. W. Roush [197].

ALGORITHM 3.2.1. Let $A \in B_n$.

Step 1. Find $B_r(A)$. A minimum norm g-inverse will exist if and only if the sets of 1's in the members of $B_r(A)$ are disjoint, i.e., for $v, w \in B_r(A)$, $v \neq w$, $v_i = 1$ implies $w_i = 0$.

Step 2. Let E be the matrix $e_{ij} = 1$ if and only if for some v $\in B_r(A)$, $v_i = v_j = 1$.

Step 3. Any solution G of GA = E will be a minimum norm g-inverse of A.

To find such solutions, for each row of E. One such solution can easily be found by adding all rows of A less than or equal to the rows of E in question. For low dimensions some sort of branch and bound method can be used if all possibilities are desired.

EXAMPLE 3.2.1. Let A be as in Example 3.1.2. Then

$$E = \begin{bmatrix} 1 & 0 & 0 \\ 0 & 1 & 1 \\ 0 & 1 & 1 \end{bmatrix}$$

So $G_{1*} = (0\ 1\ 1)$, $G_{2*} = (0\ 1\ 0)$, and $G_{3*} = (0\ 1\ 0)$. Thus

$$G = \begin{bmatrix} 0 & 0 & 1 \\ 0 & 1 & 0 \\ 0 & 1 & 0 \end{bmatrix}$$

A similar algorithm holds for A_l^- and is stated below omitting the steps as it follows on the same lines as A_m^-.

ALGORITHM 3.2.2. Let $A \in B_n$. Apply the preceding algorithm to A^T and transpose the result.

The properties of minimum norm g-inverses readily follow from the properties of symmetric idempotents, i.e., matrices of partial equivalence relations.

THEOREM 3.2.3 [197]. Let $A \in B_n$. Then A has a minimum norm g-inverse if and only if no two row basis vectors of A has a 1 in the same column. If an inverse exists, all inverses are given by Algorithm 3.2.1.

Proof. By Theorem 3.2.1, $G = A_m^-$ if and only if GA is a symmetric idempotent with the same row space as A.

By Corollary 3.1.12, the row basis vectors of this idempotent matrix E will have non-overlapping sets of ones. But E has the same row space as A since $E \in L_A$. Thus the same will be true of a row basis for A.

If A has this property, then $e_{ij} = 1$ if and only if $v_i = 1$ and $v_j = 1$ for some basis vector v of A will be a symmetric and idempotent with the same row space as A. To show E is unique if E_1 is another symmetric idempotent with the same row space as A, E_1 by symmetry must also have the same column space as the row space of A. Thus E, E_1 lie in the same H-class and so $E = E_1$. □

PROPOSITION 3.2.4. A matrix G is a least squares g-inverse of A if and only if G^T is a minimum norm g-inverse of A^T.

Proof. This follows from the defining formulas $A = AGA$, $(AG)^T = AG$. □

DEFINITION 3.2.2. The *Stirling's number of the second kind* S(n, m) is the number of equivalence relations having m equivalence clsses on $\underline{n}$.

There are formulas

$$S(n, m) = (m!)^{-1}\sum_{k=1}^{m} (-1)^{m-k}\binom{m}{k}k^n$$

and $S(n + 1, r + 1) = (r + 1)S(n, r + 1) + S(n, r)$ for the Stirling's number of the second kind.

THEOREM 3.2.5 [197]. The number of matrices of rank k, dimension n, which have a minimum norm g-inverse is

$$S(n + 1, k + 1) \sum_{i=0}^{n} (-1)^i \binom{k}{i} (2^k - i)^n$$

Proof. Choosing a row space amounts to dividing the set of n positions of vectors in V_n into either k disjoint nonempty sets or k + 1 disjoint nonempty sets with one set selected as those columns which are always zero. This can be done in $S(n, k) + (k + 1)S(n, k + 1)$ ways.

A matrix with such a row space will have a minimum norm g-inverse. By Theorem 1.3.7 there are

$$\sum_{i=0}^{k} (-1)^i \binom{k}{i} (2^k - i)^n$$

matrices with a particular choice of such row space. (This uses the fact that the space generated by k vectors with disjoint sets of 1's has 2^k elements). This completes the proof. □

3.3 Thierrin-Vagner Inverses

In this section we study the existence of Thierrin-Vagner inverses. We also give an algorithm for finding all Thierrin-Vagner inverses of square matrices.

All regular Boolean matrices have Thierrin-Vagner inverses (also known just as inverses, to semigroup theorists). Any nonsingular matrix has a unique Thierrin-Vagner inverse which can be computed by cofactors. In general, a slight modification of Algorithm 3.1.2 yields all Thierrin-Vagner inverses. The average number of Thierrin-Vagner inverses of a random regular element of B_n tends to infinity.

DEFINITION 3.3.1. Let $A \in B_{mn}$. Then $G \in B_{nm}$ is said to be a *Thierrin-Vagner inverse* of A, denoted by A* if A = AGA and G = GAG.

As we mentioned in Section 2.1 the notion of inverse was introduced in 1952 by V. V. Vagner [380] under the name of "generalized inverse" and G. Thierrin [382] who called it "reciprocal." Other authors have called it "semi-inverse," "Moore-Penrose inverse," and "reflexive g-inverse" [178, 282, 294, 296, 297]. The following is due to R. J. Plemmons [282]. However in a more general setting Theorem 3.3.1 was previously proved by G. Thierrin [382].

THEOREM 3.3.1 [282]. If $A \in B_n$ has a g-inverse, then A has a reduced Thierrin-Vagner inverse.

Proof. If A has a g-inverse, then both L_A and R_A contain idempotents. Moreover, L_A contains a row reduced idempotent E and R_A contains a column reduced idempotent F. Now according to Proposition 2.1.1 (4), A has a Thierrin-Vagner inverse G such that AG = F and GA = E. Now $G \in L_F \cap R_E$, so that G is row reduced since each member of L_F is row independent and G is column reduced since each member of R_E is column independent. That is, G is a Thierrin-Vagner inverse of A. This proves the theorem. □

The next result will show that a nonsingular matrix has a unique Thierrin-Vagner inverse.

LEMMA 3.3.2. Let $E \in B_n$ and $\rho(E) = n$. Then E is a reduced idempotent matrix if and only if (i) $e_{ii} = 1$ for every i, (ii) if $e_{ij} = 1$ for $i \neq j$, then $e_{ji} = 0$, (iii) if $e_{ij} = 1$ and $e_{jk} = 1$, then $e_{ik} = 1$.

Proof. Straightforward. □

THEOREM 3.3.3 [282]. Any nonsingular matrix $A \in B_n$ has a unique Thierrin-Vagner inverse.

Proof. Suppose A is nonsingular. Then A has a Thierrin-Vagner inverse. We shall proceed to show L_A and R_A each contain exactly one idempotent. Now each member of L_A and R_A has rank n. Moreover, L_A contains a row independent idempotent and R_A contains a column independent idempotent. All idempotents in R_A and L_A will be reduced. Suppose L_A contains two idempotents, E_1 and E_2. Then E_1 and E_2 are reduced and there exists $P \in P_n$ such that $E_2 = PE_1$. This means that $E_1 PE_1 = E_1$. We shall show that this implies $P = I$. Suppose $p_{ij} = 1$. Then $e_{ij} = 1$ since $e_{kk} = 1$ for each $1 \le k \le n$. Also for $1 \le i \le n$, $e_{ii} = 1$ implies $e_{ik} = p_{km} = e_{mi} = 1$ for some $1 \le k, m \le n$. Then $e_{km} = 1$ so that $e_{ki} = e_{ik} = 1$ by Lemma 3.3.2 (iii), $i = k$. Similarly, $i = m$, so that $p_{km} = p_{ii} = 1$. Thus $P = I$ and so $E_2 = E$. Similarly R_A contains exactly one idempotent. Then the Thierrin-Vagner inverse G of A such that AG is the idempotent in R_A and GA is the idempotent in L_A is the unique Thierrin-Vagner inverse of A. □

Notice that the converse of Theorem 3.3.3 is not, in general, true.

THEOREM 3.3.4 [296]. Let $A \in B_{mn}$. Then G is a Thierrin-Vagner inverse of A if and only if $G = R^- L^-$ where $A = LR$ is a space decomposition of A and either R^- is a Thierrin-Vagner inverse of R or L^- is a Thierrin-Vagner inverse of L.

Proof. Follows from Theorems 3.1.14, 3.1.15, and 3.1.16. □

In the following we state a semigroup-theoretic algorithm due to R. J. Plemmons [282].

ALGORITHM 3.3.1. Let $A \in B_n$.

Step 1. Find $E \in$ Idem (L_A).

Step 2. Solve the equation $XA = E$. (See Algorithm 3.2.1).

EXAMPLE 3.3.1. Consider the matrix

$$\begin{bmatrix} 1 & 0 & 1 & 1 \\ 1 & 0 & 1 & 0 \\ 1 & 1 & 1 & 0 \\ 1 & 0 & 1 & 0 \end{bmatrix}$$

First, A is row reduced to

$$A_1 = \begin{bmatrix} 1 & 0 & 1 & 1 \\ 1 & 0 & 1 & 0 \\ 1 & 1 & 1 & 0 \\ 0 & 0 & 0 & 0 \end{bmatrix}$$

By permuting the rows of A we obtain the row independent idempotent

$$E = \begin{bmatrix} 1 & 0 & 1 & 0 \\ 1 & 1 & 1 & 0 \\ 0 & 0 & 0 & 0 \\ 1 & 0 & 1 & 1 \end{bmatrix}$$

so that A has a g-inverse. Finally, solving XA = E subject to the proper restrictions we have

$$G = \begin{bmatrix} 0 & 0 & 0 & 1 \\ 0 & 0 & 1 & 1 \\ 0 & 0 & 0 & 0 \\ 1 & 0 & 0 & 1 \end{bmatrix}$$

as a reduced Thierrin-Vagner inverse of A.

DEFINITION 3.3.2. For any matrix $A \in B_n$, let $A[i|j]$ denote the matrix obtained by deleting ith row and jth column of A. The *cofactor* of a_{ij} is the permanent of $A[i|j]$. We shall call the transposed matrix of cofactors of elements of A the *adjoint* of A, and it is denoted by Adj (A).

PROPOSITION 3.3.5 [174]. If $A \in B_n$ is nonsingular, then Adj (A) $\in D_A$.

Proof. Compare with the following proof. □

THEOREM 3.3.6 [174]. If $A \in B_n$ is nonsingular, then Adj (A) = A*.

Proof. We have Adj (PAQ) = Q^T(Adj (A))P^T and (PAQ)* = $Q^T A^* P^T$ where P and Q are permutation matrices, so it suffices to prove this theorem in the case where A is a rank n idempotent in triangular form. Suppose A is such an idempotent. Then A* = A. So it will suffice to prove that Per (A[i|j]) = a_{ji}. Suppose a_{ji} = 1 and i ≠ j. Then a_{kk} for k ≠ i or j and a_{ji} give a nonzero diagonal product in A[i|j], if i ≠ j. If i = j, a_{ji} and Per (A[i|j]) both equal 1.

Suppose a_{ji} = 0 Per (A[i|j]) = 1. Let $a_{1,p(1)} a_{2,p(2)} \cdots a_{n,p(n)}$ be a nonzero diagonal product in A[i|j]. Note that the domain of $p \in \underline{n} \setminus \{i\}$ and the range of $p \in \underline{n} \setminus \{j\}$. Suppose some main diagonal entry appears in this product. Then deleting its row and column we would obtain a matrix B of lower dimension such that b_{ji} = 0 and Per (B[i|j]) = 1 for some i, j and B is in triangular form with ones on its diagonal and $B^2 \leq B$, so $B^2 = B$. Assume the dimension of A is minimal. Then by what we have just obtained, no main diagonal entries occur among the $a_{k,p(k)}$. Thus since A is triangular, p(k) < k for all k. Thus p(1) cannot be defined so i = 1 and n cannot be in the image of p so j = n. And so p maps $\underline{n} \setminus \{1\}$ onto $\underline{n-1}$ and p(k) < k for each k. So p(2) must be 1, p(3) must be 2, and general p(k) = k - 1 for each k. Thus a_{21}, $a_{32}, \ldots, a_{n,n-1}$ = 1. Thus by idempotency of A, a_{n1} = 1. But i = 1 and j = n. So a_{ji} = 1. But we assumed a_{ji} = 0. This is a contradiction.

Thus if a_{ji} = 0, Per (A[i|j]) = 0. This proves the theorem. □

In 1974, B. M. Schein [336] gave a surprising closed formula for the maximum Thierrin-Vagner inverse as follows.

THEOREM 3.3.7 [336]. If A ε Reg (B_n), then the maximum Thierrin-Vagner inverse of A is $(A^T A^C A^T)^C A (A^T A^C A^T)^C$.

Proof. Follows from Theorem 2.1.26. □

COROLLARY 3.3.8. If $A \in \text{Reg}(B_n)$, then the maximum Thierrin-Vagner inverse of A is $\bar{A}A\bar{A}$. (Recall that $\bar{A}$ is the largest subinverse of A from Definition 2.1.10).

Proof. Since $A \in \text{Reg}(B_n)$, $A\bar{A}A = A$. It follows that $A^* = \bar{A}A\bar{A}$. If G is any Thierrin-Vagner inverse of A, then $G \le \bar{A}$ and $G = GAG \le \bar{A}A\bar{A}$. □

We next present an algorithm which yields all Thierrin-Vagner inverses due to K. H. Kim and F. W. Roush [197]. This algorithm is a slight modification of Algorithm 3.1.2.

ALGORITHM 3.3.2. Let $A \in B_n$.

Step 1. Find $B_r(A)$.

Step 2. Find $I(v)$ for each $v \in B_r(A)$. A Thierrin-Vagner inverse exists if and only if $I(v)$ is nonempty for each $v \in B_r(A)$.

Step 3. Choose a vector $u \in I(v)$ for each $v \in B_r(A)$. (We will write $u(v)$ for u).

Step 4. For each chosen u, choose a vector s such that $s_i = 1$ only if $A_{i*} \le v$ and $s_i = 1$ for at least one i such that $A_{i*} = v$. (We will sometimes write $s(v)$ for s).

Step 5. If it is not true that $v^* \ge v$ implies $s(v^*) \ge s(v)$ redefine $s(v)$ to be $s_0(v) + \sum_{z<v} s_0(v)$ where $s_0(v)$ is the $s(v)$ obtained in Step 4.

Step 6. For a vector t which has only one 1 and if t is not less than any row basis vector, send t to an arbitrary sum of the vectors $s(v)$ or 0. Otherwise let $p(t)$ be the infimum of the row vectors which are greater than or equal to t, taken in the row space of A regarded as a lattice. Choose a set S_v of row basis vectors whose sum is less than or equal to $p(t)$ and let t be sent to $b = \sum_{z \in S_v} s(z)$ or if S_v is empty to 0. Here we assume t is not equal to any u.

Step 7. Linearly order all vectors with exactly one 1 in the order in which they occur as rows of the identity matrix. (Write the vectors s and q in the order of the t's and u's).

The main difference of this from Algorithm 3.1.2 is that we have $s(v^*) \geq s(v)$ if $v^* \geq v$. Also vectors t must be sent to vectors which are linear combinations of the vectors s(v).

EXAMPLE 3.3.1. Let

$$A = \begin{bmatrix} 1 & 1 & 0 \\ 1 & 1 & 1 \\ 0 & 1 & 1 \end{bmatrix}$$

Step 1. $B_r(A) = \{(1\ 1\ 0),\ (0\ 1\ 1)\}$.

Step 2. $I(1\ 1\ 0) = \{(1\ 0\ 0)\}$, $I(0\ 1\ 1) = \{(0\ 0\ 1)\}$.

Step 3. (Not needed).

Step 4. For (1 1 0), s = (1 0 0). For (0 1 1), s = (0 0 1).

Step 5. (Not needed).

Step 6. For t = (0 1 0), p(t) = 0. Thus q = 0.

Step 7. Thus

$$A^* = \begin{bmatrix} 1 & 0 & 0 \\ 0 & 0 & 0 \\ 0 & 0 & 1 \end{bmatrix}$$

REMARK 3.3.1. If it is only desired to find a single Thierrin-Vagner inverse, this procedure can be considerably simplified. We could omit Step 3, take s to be the vector $s_i = 1$ if and only if $A_{i*} \leq v$, send all identification vectors for v to this s and all non-identification vectors containing only one 1 to zero.

THEOREM 3.3.9 [197]. Let $A \in B_n$. Then Algorithm 3.3.2, for all allowable choices, yields all Thierrin-Vagner inverses of A.

Proof. First we will show all matrices G constructed by this algorithm are Thierrin-Vagner inverses of A.

First we note that the definition of s(v) implies s(v)A = v. Also for $t \leq v$, $tGA = \sum_{S_v} s(z)A = \sum_{S_v} z \leq p(t) \leq v$. Thus for a basis vector v, $vGA = s(v)A + \sum_{t \leq v} \sum_{S_v} s(z)A = v$, and so as before any vector v_2, $v_2AGA = v_2A$ and AGA = A. The image of G is generated by vectors s(v). We have s(v)A = v and $vG = s(v) + \sum_{t \leq v} \sum_{S_v} s(z)$. Each z in the second part of the sum is less than or equal to p(t) and so less than or equal to v. Thus by Step 5, $s(z) \leq s(v)$. Thus vG = s(v). So s(v)AG = s(v), and AG is the identity on the image of G. It follows that GAG = G.

Next suppose G = A*. Since G will also be a g-inverse of A, there must be identification vectors u mapped to vectors s(v) as described in Step 4. By the construction s(v)A = v. Also vGA = v, since v is in the image of A. Thus s(v) = uG and vG are vectors in the image of G which have the same image under A. But GAG = G implies A is one-to-one on the image of G. Thus s(v) = vG. Thus for $v \leq v^*$, $s(v) \leq s(v^*)$.

Let K be the subspace spanned by the vectors s(v). We have s(v)A = uGA = v so K maps onto the space of row vectors of A, under A. Also K is contained in the image of G, on which A is one-to-one.

Let t be a vector with exactly one 1 and let $W = \{w \in R(A): w \geq t\}$. For $w \in W$, wGA = w, so $tGA \leq w$. Since tGA is in the row space of A, $tGA \leq p(t)$. (Similarly for $u \leq v$). Suppose $tGA = \sum_{S_v} z$. Since $\sum_{S_v} s(z)A = \sum_{S_v} z$ and A is one-to-one on the image of G, tG must be $\sum_{S_v} s(z)$. If $W = \emptyset$, tG is in the image of G which in turn equals K. This proves the theorem. □

REMARK 3.3.2. The number of Thierrin-Vagner inverses of A in D_A is equal to the product of the number of idempotents in L_A and the number of idempotents in R_A.

PROPOSITION 3.3.10 [206]. The average number of Thierrin-Vagner inverses of elements of Reg (B_n) tends to infinity as n increases.

Proof. Exercise.

We next mention some results about Hamming distances in relation to Thierrin-Vagner inverses. The Thierrin-Vagner inverses were proposed as a method of encoding data, for instance pictorial data by J. Denes. The results show that this would not be a good method (another disadvantage of this method is that a random $n \times n$ Boolean matrix for n large has a very small probability of being regular).

DEFINITION 3.3.3. If $A, B \in B_n$ the *Hamming distance* $d(A, B) = |\{i, j\}: a_{ij} \neq b_{ij}|$. The *weight* $w(A) = d(A, 0)$.

In K. H. Kim and F. W. Roush [201] they considered a number of problems of minimum Hamming distance and also average weights of various products. For instance $d(BA, I)$ is minimized for a given A if and only if (i) $B_{r*} = I_{s*}$ if $A_{r*} = I_{s*}$, (ii) if $w(A_{r*}) = 2$ and $a_{rr} = 1$, then $B_{r*} = A_{r*}$ or $B_{r*} = 0$, and (iii) $B_{r*} = 0$ for all other r. Thus the Vagner inverse is not very good in this sense. We also proved

$$\sum_{X \in B_n} w(AX) + \sum_{X \in B_n} d(BAX, X) \geq n^n 2^{n^2-1}$$

with equality only if A is a partial permutation matrix. The average weight of a product of two random $n \times n$ Boolean matrices is

$$n^2\left(1 - \left(\frac{3}{4}\right)^n\right)$$

since $1 - (\frac{3}{4})^n$ is the probability that any particular entry is 1. Moreover, they studied some more combinatorial problems relating to weights of Boolean vectors.

3.4 Moore-Penrose Inverses

We shall present necessary and sufficient conditions for a matrix to possess Moore-Penrose inverses. We shall also give an algorithm for finding all Moore-Penrose inverses.

A matrix G is a Moore-Penrose inverse of A if and only if it is a minimum norm g-inverse and a least squares g-inverse. For matrices over the real numbers, the Moore-Penrose inverse has proved quite useful. In fact every matrix over the real numbers has a unique Moore-Penrose inverse.

Only a restricted class of Boolean matrices has Moore-Penrose inverses, but this class, the so-called *difunctional matrices*, is of special interest. Moreover if a Moore-Penrose inverse exists it is unique, and in fact must be the transpose of the original matrix.

DEFINITION 3.4.1. Let $A \in B_{mn}$. Then $G \in B_{nm}$ is said to be a *Moore-Penrose inverse* of A, denoted A^+ if $A = AGA$, $G = GAG$, $AG = (AG)^T$, and $GA = (GA)^T$.

We shall begin with the following result due to P. P. Rao [296].

THEOREM 3.4.1 [296]. Let $A \in B_{mn}$. Then the following statements are equivalent. (i) A^+ exists. (ii) A is of the form

$$P\begin{bmatrix} I & C \\ D & DC \end{bmatrix}Q$$

where P and Q are permutation matrices and C and D are such that $CC^T \le I$ and $D^TD \le I$. (iii) A^- exists, $C(A) = C(AA^T)$ and $C(A^T) = C(A^TA)$. (iv) There exists G such that $GAA^T = A^T$ and $A^TAG = A^T$.

Proof. Follows immediately from Theorem 3.2.1 and 3.2.2. □

The following corollary was obtained independently by J. Riguet [311], C. R. Rao [294], and R. J. Plemmons [282].

COROLLARY 3.4.2. Let $A \in B_n$. If A^+ exists, then $A^+ = A^T$.

DEFINITION 3.4.2. Let $A \in B_n$. Then A is said to be *difunctional* if $A^+ = A^T$, i.e., $AA^TA = A$.

The difunctional relations have been extensively studied by J. Riguet [311-313], and B. M. Schein [337].

The following result is well known. As a consequence of it, for any difunctional Boolean matrix A there exist permutation matrices P and Q such that PAQ can be partitioned into blocks each of which is either 0 or J such that no block row or column contains more than one J-block.

THEOREM 3.4.3. Let $A \in B_n$. Then A has a Moore-Penrose inverse if and only if any two rows of A are either equal or have disjoint sets of 1's.

Proof. Necessity. By Theorem 3.2.3, the set of 1's in any two row basis vectors are disjoint and so that the set of 1's in any two column basis vectors are disjoint. Suppose some row A_{c*} is a nontrivial sum involving two row vectors A_{a*}, A_{b*} which are different basis vectors. There will be some i, j such that $(A_{a*})_i = 1$, $(A_{b*})_i = 0$, $(A_{a*})_j = 0$, $(A_{b*})_j = 1$. Thus there will be vectors u, v in the column basis such that $u_a = 1$, $u_b = 0$, $v_a = 0$, $v_b = 1$. However all column vectors which are 1 in either the ath row or the bth row will have a 1 in the cth row, and so u, v do not have disjoint sets of 1's, a contradiction.

Sufficiency. By Corollary 3.4.2, if A has a Moore-Penrose inverse, then $A^+ = A^T$. We will show that A^T is a Moore-Penrose inverse of A, if A satisfies the condition of this theorem. The (i, j)-entry of AA^TA is 1 if and only if there exist p, q such that $a_{ip} = (A^T)_{pq} = a_{qj} = 1$, or $a_{ip} = a_{qp} = a_{qj} = 1$. Thus the ith row and the qth row are not disjoint, so they must be equal so that $a_{ij} = 1$. Conversely if $a_{ij} = 1$, taking p = j, q = i gives $(AA^TA)_{ij} = 1$. Thus $AA^TA = A$. Transposing this equation, we obtain $A^TAA^T = A^T$. The matrices AA^T

and A^TA will always be symmetric. This proves the theorem. □

PROPOSITION 3.4.4 [197]. The number of Boolean matrices of order n with rank k having a Moore-Penrose inverse is $k!(S(n, k) + (k + 1)S(n, k + 1))^2$ where the $S(n, k)$ are Stirling numbers of the second kind.

Proof. Exercise.

In conclusion, we remark that the theory of nonnegative inverse of nonnegative matrices over the real number is very similar to the preceding theory of Boolean inverses, and some cases the Boolean theory can used to study the nonnegative case [282, 296]. One difference is that every full rank idempotent nonnegative matrix is a diagonal matrix. Thus in various theorems, diagonal matrices play the role that full rank idempotent matrices play in the Boolean case.

The other special inverses were studied by J. Denes [75, 77], J. Denes, K. H. Kim and F. W. Roush [84], and M. P. Drazin [93].

Exercises

1. Find the g-inverses, minimum norm g-inverse, least squares g inverse, Thierrin-Vagner inverse, and Moore-Penrose inverse of the following matrices, or show that an inverse does not exist:

$$\begin{bmatrix} 1 & 0 & 1 \\ 0 & 1 & 1 \\ 1 & 1 & 1 \end{bmatrix}, \quad \begin{bmatrix} 1 & 1 & 0 & 0 \\ 0 & 1 & 1 & 0 \\ 0 & 0 & 0 & 1 \\ 0 & 1 & 1 & 0 \end{bmatrix}, \quad \begin{bmatrix} 1 & 0 & 1 & 0 & 1 \\ 1 & 0 & 1 & 1 & 0 \\ 0 & 0 & 0 & 0 & 1 \\ 1 & 1 & 1 & 0 & 1 \\ 0 & 1 & 0 & 0 & 1 \end{bmatrix}$$

2. Prove that if M is a fully indecomposable Boolean matrix then M is a sum of permutation matrices and thus has permanent at least 2, unless $M \in B_1$. To do this note that if we delete the row and column of any entry of M we obtain a Hall matrix. Thus M has a permutation passing through each 1 entry.

3. Use Exercise 2 to show that a (0, 1)-matrix M has permanent 1 if and only if there exist permutations P, Q such that PMQ is reflexive and subtriangular. Either M is fully indecomposable or PMQ will have the form

$$\begin{bmatrix} * & 0 \\ & \\ * & * \end{bmatrix}$$

for some P, Q. Then repeat this construction on the diagonal blocks.

4. Prove that if M is subtriangular and reflexive $\{j: A_{j*} \le A_{i*}\} \subset \{j: a_{ij} = 1\}$. This with Exercise 3 can be used to prove Rao's criterion for existence of g-inverse.

5. Prove that the average number of inverses of an element x in regular D-class is

$$\frac{|\text{Idem }(D_x)|\,|H_x|}{|D_x|}$$

Prove that for D-classes of rank n - r, this number tends to infinity as min {r, n - r} does.

6. Prove Corollary 3.1.3.
7. Prove Propositio-n 3.1.6.
8. Prove Corollary 3.1.8.
9. Prove Corollary 3.1.11.
10. Prove Corollary 3.1.12.
11. Prove Proposition 3.1.16.
12. Prove Proposition 3.1.17.
13. Prove Corollary 3.1.18.
14. Prove Corollary 3.1.19.
15. Prove Corollary 3.1.20.
16. Prove Proposition 3.1.21.
17. Prove Corollary 3.1.22.
18. Prove Corollary 3.1.25.
19. Prove Corollary 3.1.26.

20. Prove Corollary 3.1.28.

21. Prove Proposition 3.3.5.

22. Prove Proposition 3.3.10.

23. Prove Proposition 3.4.4.

24. Prove

$$\sum_{x \in B_n} w(AX) = n2^{n^2-n} \sum_{j=1}^{n} (2^n - 2^{n-w_j}), \; w_j = w(A_{j*})$$

To do this, show that the average weight of $A_{j*}X_{*k}$ is $1 - 2^{-w_j}$. (This weight is 0 or 1).

25. Let A, B, C $\in B_n$. What is the average weight of a product ABC ? Define a Markov process related to the average weight of a k-fold product of Boolean matrices ?

26. Draw the graph of the transformation f such that f(1) = 2, f(2) = 3, f(3) = 4, f(4) = 5, f(5) = 2.

27. Draw the graph of the transformation f such that f(1) = 2, f(2) = 3, f(3) = 4, f(4) = 4.

28. Draw the graph of the transformation f such that f(1) = f(2) = 3, f(3) = 4.

29. Draw the graph of the transformation f such that f(1) = 2, f(2) = 3, f(3) = 2, f(4) = 5, f(5) = 6, f(6) = 7, f(7) = 5.

30. Write the Boolean matrices of transformations in Exercise 26-29.

31. Find Thierrin-Vagner inverses of the transformations in Exercise 26-29.

32. Prove that all nonnegative idempotent matrices A over $\mathcal{R}$ have the form $P^{-1}BP$ where B is a direct sum of zero matrices and rank 1 matrices and P is a permutation matrix.

33. Prove that if A is a nonnegative matrix over $\mathcal{R}$ having a nonnegative Thierrin-Vagner inverse that PAQ is a diagonal matrix where P and Q are permutation matrices.

34. What are all symmetric idempotent matrices over the nonnegative real numbers ?

35. Prove a nonnegative matrix A has a nonnegative least squares g-inverse if and only if there exist k nonzero columns in which

the locations of the nonzero entries are distinct where k is the rank of A.

36. Find a least squares g-inverse of this matrix.

$$\begin{bmatrix} 1 & 0 & 3 \\ 0 & 2 & 2 \\ 0 & 3 & 3 \end{bmatrix}$$

(Find a symmetric idempotent E having the same column space. Then solve AX = E).

37. Prove that if a nonnegative matrix A has a nonnegative Moore-Penrose inverse then any two rows either have inner product 0, or one is a multiple of the other. Find a formula for the Moore-Penrose inverse if this condition holds.

CHAPTER FOUR

COMBINATORIAL PROPERTIES OF ORDER RELATIONS

4.1 Topologies

Since nonsingular idempotent Boolean matrices are the matrices of partial order relations, the theory of partially ordered sets (posets) is important in Boolean matrix theory. It turns out that on finite sets, quasi-order structures are equivalent to topologies.

Finite topologies have been studied by a number of authors, and we recount the history of the subject. One still unsolved problems (for $n > 11$) is to count the number of topologies on $\underline{n}$. S. K. Das [60] has calculated this number for $n > 11$. D. J. Kleitman and B. L. Rothschild [215] obtained the asymptotic number of topologies on $\underline{n}$, by showing that most posets have a very simple form consisting of three stages (their proof is quite intricate and involves consideration of many separate cases). Here we simplify the Kleitman-Rothschild result to

$$\sqrt{\frac{2}{\pi n}}\, c_i 2^{\frac{(n+1)^2}{4} + n}$$

for constants c_1, c_2.

We also give the asymptotic number of quasi-orders on $\underline{n}$, of transitive relations on $\underline{n}$, the log asymptotic number of regular $n \times n$ Boolean matrices, and estimate the number of $n \times n$ idempotent Boolean matrices.

DEFINITION 4.1.1. A *topology* on a finite set S is a family F of subsets of S such that (i) $\emptyset \in F$, (ii) $S \in F$, (iii) for all $X, Y \in F$ both $X \cup Y \in F$ and $X \cap Y \in F$. The elements of F are called *open sets*. A set is *closed* if and only if its complement is open. The closed sets also form a (different) topology. For a set X, X' is the intersection of all closed sets containing X. Here X' denotes the closure of X.

For any topology, the relation $x' \subset y'$ is a partial order. Conversely for any quasi-order Q the family of all unions of sets $U_a = \{x: (x, a) \in Q\}$ is a topology. This gives a one-to-one correspondence between all topologies on a finite set and all quasi-orders on the set, and between all T_0-topologies (that is for all x, y one of x, y lies in an open set not containing the other) and all partial orders. These correspondences are due to P. S. Alexandrov [7] and G. Birkhoff [22], R. L. Blair [26], W. J. Thron [383], H. J. Kowalski [218], H. Sharp [361], K. H. Kim and W. Maryland [185] obtained other results relating topologies and order structures.

M. C. McCord [248] and R. E. Stong [377] showed that finite topological spaces have exactly the same sets of homotopy and homology groups as finite polyhedra, so that algebraic topology can be used. Several authors have proved fixed-point theorems for partially ordered sets. J. Neggers [265], K. Johnson [165], R. P. Stanley [374], I. Rabinovitch [293], K. H. Kim [175], and W. Maryland [246] obtained other structural results about partially ordered sets.

V. Krishnamurty [222], J. W. Evans, F. Harary and M. S. Lynn [112], J. Wright [404], M. Erne [111], and S. K. Das [60] have used computers to enumerate partially ordered sets.

The asymptotic number of partially order structures on $\underline{n}$ has been studied by S. D. Chatterji [51], L. Comtet [58], D. J. Kleitman and B. L. Rothschild [214, 215]. Formulas for counting special kinds of partially ordered sets and other structures were found by D. Klarner [211], R. P. Stanley [373], K. H. Kim [177], K. H. Kim and G. Markowsky [183, 184], and R. L. Davis [66].

Other results on partially ordered sets and finite topologies were obtained by S. S. Anderson and G. Chartrand [10], K. H. Kim [176], J. Neggers [262-264], and K. H. Kim and F. W. Roush [191].

DEFINITION 4.1.2. (i) Let p_n denote the number of partial order relations on $\underline{n}$. (ii) Let q_n denote the number of quasi-order relations on $\underline{n}$. (iii) Let t_n denote the number of transitive relations on $\underline{n}$. (iv) Let r_n denote the number of reflexive relations on $\underline{n}$.

THEOREM 4.1.1 [112]. p_n and q_n are asymptotically equal.

Proof. It is well known that

$$q_n = \sum_{m=1}^{n} p_m S(n, m)$$

From the equation $S(n + 1, n - k) - S(n, n - k - 1) = (n - k)S(n, n - k)$ we can derive the inequality $S(n, n - k) \le n^{2k}$, by induction on k.

D. J. Kleitman and B. L. Rothschild [214] showed that

$$\log p_n = \frac{n^2}{4} + \frac{3n}{2} + O(\log n)$$

Thus there exists a constant c such that

$$\frac{q_n - p_n}{p_n} < \sum_{k=1}^{n-1} 2^{-\frac{kn}{2} + \frac{k^2}{4} + c(\log n) + 2k(\log n)}$$

for all sufficiently large n. Take n large enough that

$$\frac{n}{\log n} \ge 10$$

Then n times any term of the expression above is less than or equal to

$$2^{-\frac{kn}{20} + (c+1)\log n}$$

For any sequence of choices of k, this will tend to 0 as n tends to infinity. Thus

$$\frac{q_n - p_n}{p_n}$$

tends to 0. This proves the theorem. □

THEOREM 4.1.2 [206]. t_n is asymptotically equal to $2^n p_n$.

Proof. We define a mapping from the set of transitive relations to the set of quasi-order relations. Let T be the matrix of a transitive relation. Then $T^2 \leq T$. Thus $I + T$ is idempotent, and is the matrix of a unique quasi-order relation. Under this correspondence at most 2^n transitive relations correspond to any quasi-order relation. Thus $t_n \leq 2^n q_n$.

On the other hand from any partial order we can obtain 2^n transitive relations by changing ones on the diagonal to zeros. These will be distinct for different partial orders. Thus $t_n \geq 2^n p_n$. Now this theorem follows from Theorem 4.1.1. □

THEOREM 4.1.3 [206]. $\text{Log } |\text{Idem } (B_n)| \geq \frac{n^2}{4} + 2n + 0(\log n)$.

Proof. Consider all height 3 graded posets with exactly $[\frac{n}{4}]$ elements on the top level and exactly $[\frac{n}{4}]$ elements on the bottom level. There are

$$2^{\frac{n^2}{4} + \frac{3n}{2} + 0(\log n)}$$

of these. However we can obtain at least $2^{n/2}$ idempotents for each of these by entering diagonal zeros in locations corresponding to a subset of the elements in the middle level. All these idempotents will be distinct. □

THEOREM 4.1.4 [206]. $\text{Log } r_n = \frac{n^2}{2} + 0(n(\log n))$

Proof. The number of regular matrices is less than or equal to n! times the number of H-classes of regular matrices, since no H-class of B_n has more than n! elements. Each H-class is determined by the R-class it lies in and the L-class it lies in. Each R-class and each L-class of regular matrices contains an idempotent, and any such idempotent determines the class. Thus the number of such pairs of R, L-classes is less than or equal to the number of pairs of idempotents which by Theorem 4.1.2 is

$$2^{\left(\frac{n^2}{4} + 0(n)\right)2}$$

Thus

$$|\text{Reg } (B_n)| \leq n!2^{\frac{n^2}{2} + 0(n)}$$

For the lower bound we consider the number of matrices in the regular D-class containing all rank $[\frac{n}{2}]$ partial permutation matrices. The size of any D-class equals

$$\frac{(\text{cardinality of R-class})(\text{cardinality of L-class})}{(\text{cardinality of H-class})}$$

Here an H-class has $[\frac{n}{2}]!$ elements. Since the row rank of these matrices equals their column rank, it follows from a result of R. L. Brandon, D. W. Hardy, K. H. Kim and G. Markowsky [40] that the cardinality of an L-class equals that of an R-class. The R-class of

$$\left[\begin{array}{c|c} I & 0 \\ \hline 0 & 0 \end{array}\right]$$

contains all matrices of the form

$$\left[\begin{array}{c|c} I & * \\ \hline 0 & 0 \end{array}\right]$$

Thus the size of D-class is at least

$$\left(\left[\frac{n}{2}\right]!\right)^{-1} 2^{\left(\frac{n^2 - 1}{4}\right)2}$$

□

We now show that the D. J. Kleitman and B. L. Rothschild [214] asymptotic formula for the number of posets on $\underline{n}$ can be simplified somewhat. The following results are due to K. H. Kim and F. W. Roush [203].

D. J. Kleitman and B. L. Rotschild [214] showed that the number of posets on $\underline{n}$ is asymptotically equal to

$$\sum_{\substack{a+b+c=n\\ b\geq 1, c\geq 1}} \binom{n}{a,b,c}(2^b - 1)^a (2^c - 1)^b \tag{4.1.1}$$

We will attempt to simplify this result.

We will compare (4.1.1) with the simpler sum

$$\sum_{a+b+c=n} \binom{n}{a,b,c} 2^{ab} 2^{bc} \tag{4.1.2}$$

We need the following lemmas.

LEMMA 4.1.5. Formula (4.1.2) is asymptotically equal to the sum of these of its terms for which a, b, c are all at least n/8.

Proof. Write Formula (4.1.2) as

$$\sum_{a+b+c=n} \binom{n}{b}\binom{a+c}{a} 2^{b(n-b)}$$

For $a = c = [\frac{n}{4}]$, $b = n - 2[\frac{n}{4}]$, the term of the sum will be at least (for large n)

$$\frac{\sqrt{8}}{\pi n}\, 2^{\frac{n^2}{4} + \frac{3n}{2} - 4} \tag{4.1.3}$$

Here the formula

$$\binom{2k}{k} \sim \frac{2^{2k}}{\sqrt{\pi k}}$$

was used for each binomial coefficient.

Since there are less than n^2 terms in the sum, we can disregard any term which is asymptotically no more than $1/n^3$ times Formula (4.1.3). Since the multinomial coefficient is at most 3^n, for large n all terms with n/2, all terms with $|\frac{n}{2} - b| > \sqrt{n(\log n)}$ can be disregarded.

Suppose $|\frac{n}{2} - b| \le \sqrt{n(\log n)}$. Then

$$n^3 \frac{\binom{n}{b}\binom{n-b}{a}2^{b(n-b)}}{\binom{n}{b}\binom{n-b}{b'}2^{b(n-b)}}, \quad b' = [\frac{n-b}{2}]$$

will tend to 0 for $a < \frac{n}{8}$ or for $c = n - b - a < \frac{n}{8}$ by Stirling's formula. Thus terms of this type can be ignored. □

LEMMA 4.1.6. The sums (4.1.1) and (4.1.2) are asymptotically equal.

Proof. Sum (4.1.2) is greater than (4.1.1), so we need only prove (4.1.2) is asymptotically less than or equal to sum (4.1.1). By Lemma 4.1.5 it will suffice to show that there exists a function f(n) such that $f(n) \to 1$ and

$$\frac{2^{ab}2^{bc}}{(2^b - 1)^a(2^c - 1)^b} \le f(n)$$

whenever a, b, $c \ge \frac{n}{8}$. Thus we need to consider $(1 - 2^{-b})^{-a}(1 - 2^{-c})^{-b}$. This is at most

$$(1 - 2^{-n/8})^{-n}(1 - 2^{-n/8})^{-n}$$

This expression tends to 1, so take it as f(n). □

LEMMA 4.1.7. Sum (4.1.2) equals

$$\sum_{b=0}^{n} \binom{n}{b} 2^{(b+1)(n-b)}$$

Proof. Write (4.1.2) as

$$\sum_{a+b+c=n} \binom{n}{b}\binom{n-b}{a}2^{b(n-b)}$$

$$= \sum_{b=0}^{n} \binom{n}{b} 2^{b(n-b)} \sum_{a=0}^{n-b} \binom{n-b}{a}$$

and perform the second summation. This proves the lemma. □

THEOREM 4.1.8 [203]. p_n is asymptotically

$$\sqrt{\frac{2}{\pi n}}\, c_i\, 2^{\left(\frac{n+1}{4}\right)^2 + n}$$

where

$$c_1 = \sum_{n=-\infty}^{\infty} 2^{-n^2},\quad c_2 = \sum_{n=-\infty}^{\infty} 2^{-\left(\frac{n+1}{2}\right)^2},\quad i \equiv n \pmod 2.$$

Proof. We need to show

$$\sum_{b=0}^{n} \binom{n}{b} 2^{(b+1)(n-b)} \tag{4.1.4}$$

is asymptotically equal to the given expression. The term of (4.1.4) for $b = [\frac{n-1}{2}]$ is at least

$$\sqrt{\frac{2}{\pi n}}\, 2^{\frac{(n+1)^2}{4} + n - 2} \tag{4.1.5}$$

for large n, where we have again used

$$\binom{2k}{k} \sim \frac{2^{2k}}{\sqrt{\pi k}}$$

on the binomial coefficient. Since there are n + 1 terms, any set of terms in the sum which is no more than

$$\frac{1}{(n+1)^2}$$

times (4.1.5) for large n, can be disregarded without affecting the asymptotic value.

Let $k = \frac{n-1}{2} - b$. Then (4.1.4) is

$$\sum_{b=1}^{n} \binom{n}{b} 2^{\frac{(n+1)^2}{4} - k^2}$$

If $|k| > \log n$, then the terms of the above expression will be no more than $1/n^3$ times (4.1.5) for large n. Thus (4.1.4) is asymptotically equal to

$$2^{\frac{(n+1)^2}{4}} \binom{n}{[\frac{n}{2}]} \sum_{k=-t}^{t} 2^{-k^2} \frac{\binom{n}{\frac{n-1}{2}-k}}{\binom{n}{[\frac{n}{2}]}}, \quad t = \log n$$

where the summation is over all k with $|k| \leq \log n$. The numbers

$$\frac{\binom{n}{\frac{n-1}{2}-k}}{\binom{n}{[\frac{n}{2}]}}$$

will uniformly approach 1 for $|k| \leq \log n$ by the normal approximation to the binomial distribution. Thus (4.1.4) is asymptotically equal to

$$2^{\frac{(n+1)^2}{4}} \binom{n}{[\frac{n}{2}]} \sum_{k=-t}^{t} 2^{-k^2}, \quad t = \log n$$

over k with $|k| \leq \log n$. But this in turn equals asymptotically

$$2^{\frac{(n+1)^2}{4}} \binom{n}{[\frac{n}{2}]} \sum_{k=-\infty}^{\infty} 2^{-k^2}$$

And this quantity is asymptotically equal to the formula given in the theorem. This proves the theorem. □

4.2 Similarity Relations on Partially Ordered Sets

A similarity relation is a relation which is connected to a partial order in a certain way. A similarity relation is a reflexive, symmetric binary relation R such that if x R y and z is between x and y in the partial order then x R z and z R y. The matrix of a similarity order on the usual order on $\underline{n}$ is filled in towards the main diagonal. For instance this Boolean matrix;

$$\begin{bmatrix} 1 & 1 & 1 & 0 \\ 1 & 1 & 1 & 0 \\ 1 & 1 & 1 & 1 \\ 0 & 0 & 1 & 1 \end{bmatrix}$$

is the matrix of a similarity relation. Similarity relations on finite linear ordered sets are well understood. But on a product of two linearly ordered sets, little is known. Here we give estimates dealing with the number of semiorders on a partially ordered set which is a product.

DEFINITION 4.2.1. Let P be a finite poset, and let x P y denote the relation $x < y$ in the partial order. Then a *similarity relation* on P is a reflexive, symmetric binary relation R such that if x R y, x P y, x P z, z P y, then x R z and z R y.

D. G. Rogers [318] showed that the number of similarity relations on a linearly ordered set of m elements is C_m, the mth *Catalan number*. Incidentally,

$$C_m = \frac{1}{m+1}\binom{2m}{m}$$

We consider the number of similarity relations in an arbitrary finite poset and show that this can be interpreted as the number of antichains on a related poset. These results are due to K. H. Kim and F. W. Roush [206].

In addition we give estimates for the number of similarity relations on the posets V_m of all m component Boolean vectors and posets which are the product of a fixed poset and a linearly ordered set.

DEFINITION 4.2.2. Let $<_L$ be a linear order on the underlying set P which refines the partial order of P. We define a poset $S_q(P)$ as follows: the elements of $S_q(P)$ are all ordered pairs $(x, y) \in P$ such that $x <_L y$. We say that $(x, y) \leq (w, z)$ if and only if $w \hat{P} x$, $x \hat{P} y$,

$y \hat{P} z$ where $\hat{P}$ is the relation less than or equal to in the partial order of P. (It can be checked that this relation is reflexive, antisymmetric, and transitive).

We next give a definition of $S_q(P)$ which does not depend on the linear order $<_L$.

DEFINITION 4.2.3. The elements of $S_q(P)$ are all unordered pairs (x, y) such that $x \neq y$. We say that $(x, y) \leq (w, z)$ if and only if one of the following holds: (i) $w \hat{P} x$, $x \hat{P} y$, $y \hat{P} z$, (ii) $w \hat{P} y$, $y \hat{P} x$, $x \hat{P} z$, (iii) $z \hat{P} x$, $x \hat{P} y$, $y \hat{P} w$, (iv) $z \hat{P} y$, $y \hat{P} z$, $x \hat{P} w$.

DEFINITION 4.2.4. An *ideal in a poset* is a set K such that $x \in K$, $y \leq x$ imply $y \in K$.

DEFINITION 4.2.5. An *antichain in a poset* is a set C such that $x \not< y$ for any $x, y \in C$.

THEOREM 4.2.1 [206]. Let P be a poset and let $<_L$ be a refinement of the partial order on P to a linear order. Then for any similarity relation R on P the set of ordered pairs (x, y) such that $x <_L y$ and $x R y$ is an ideal in $S_q(P)$. Conversely let K be any ideal in $S_q(P)$ and let R be the binary relation such that $x R y$ if and only if $x = y$ or $(x, y) \in K$ or $(y, x) \in K$. Then R is a similarity relation. These two mappings establish a one-to-one correspondence between similarity relations on P and ideals in $S_q(P)$.

Proof. The first assertion follows from the definition of a similarity relation. Let K be an ideal in $S_q(P)$. The relation R defined as in the theorem will be reflexive and symmetric. Let $x R y$, $x P y$, $x P z$, $z P y$.

Then $x <_L y$ so $(x, y) \in S_q(P)$. Likewise $(x, z), (z, y) \in S_q(P)$. And in $S_q(P)$, $(x, y) > (x, z)$ and $(z, y) < (x, y)$. Thus $(x, z), (z, y) \in K$. Thus $x R z$ and $z R y$.

The two mappings given in the theorem are inverse to each other so establish a one-to-one correspondence. □

We remark that there is a one-to-one correspondence between ideals and antichains on a poset given by taking the maximal elements of an ideal.

COROLLARY 4.2.2. The number of similarity relations on a poset P equals the number of antichains on $S_q(P)$.

We can divide $S_q(P)$ into two parts.

DEFINITION 4.2.6. Let S_1 be the subset of $S_q(P)$ consisting of all pairs (x, y) such that x P y is false. Let S_2 be the subset of $S_q(P)$ consisting of all pairs (x, y) such that x P y. The elements of S_1 and S_2 are called *incomparable* and *comparable pairs*, respect-ively. No element of S_1 is either greater than or less than any element of $S_q(P)$.

The *cardinal sum* of two posets is the natural notion of disjoint union of the posets. Thus $S_q(P) = S_1 + S_2$ where S_1, S_2 are regarded as posets by inheriting the partial order from P.

COROLLARY 4.2.3. Let S be the set of incomparable elements in a poset P. Let P' denote the poset obtained from P by reversing the order relation. Let T denote the partially ordered subset of $P' \times P$ consisting of all pairs (a, b) such that a P b. Then the number of similarity relations on P is $2^{|S|}$ times the number of antichains on T.

Proof. The number of antichains in a cardinal sum of posets is the product of the numbers of antichains in the summand posets. It can be checked that T really is a partially ordered subset of $P' \times P$. □

REMARK 4.2.1. The set V_m can be partially ordered by setting $(x_1, \ldots, x_m) \le (y_1, \ldots, y_m)$ if and only if $x_i \le y_i$ for each i.

PROPOSITION 4.2.4. The number of incomparable pairs of V_m is

$$2^{2m-1} - 3^m + 2^{m-1}$$

Proof. The number of pairs (u, v) such that $u \le v$ is 3^m. The number of pairs (u, v) such that $u = v$ is 2^m. Thus the number of comparable pairs (u, v) is $3^m - 2^m$. The number of all pairs in V_m is

$$\binom{2^m}{2}$$

□

PROPOSITION 4.2.5. Let S denote the poset {a, b, 1} such that $x > y$ in S if and only if $x = 1$ and $y = a$ or $y = b$. Then the partially ordered subset of $S_q(V_m)$ consisting of comparable pairs is isomorphic to the partially ordered subset of S^m consisting of m-tuples with at least one 1 entry.

Proof. The partially ordered subset of $S_q(V_m)$ consisting of comparable pairs has elements all pairs of vectors (u, v) such that $u < v$ and has the following order relation: $(u', v') \le (u, v)$ if and only if $u \le u' \le v' \le v$. This is a partially ordered subset of the poset of pairs of vectors (u, v) such that $u \le v$ and $(u', v') \le (u, v)$ if and only if $u \le u' \le v' \le v$. The latter poset is the mth power of S. And 1 entries in S represent such that $u_i = 0$, $v_i = 0$. Thus there will be no 1 entries if and only if $u = v$. □

PROPOSITION 4.2.6. The number of similarity relations on V_m is at least

$$2^{2^{2m-1}} - 3^m + 2^{m-1}\binom{m}{[\frac{m}{3}]}2^{m-[\frac{m}{3}]}$$

Proof. The first three terms come from Corollary 4.2.3 and Proposition 4.2.4. The subset of S^m consisting of all m-tuples with exactly $[\frac{m}{3}]$ ones is an antichain in S^m lying in the partially ordered subset of m-tuples with at least one 1 entry. Any subset of it gives an antichain in T, where T is as in Corollary 4.2.3. This gives the last term. □

REMARK 4.2.2. The poset S^m has a large symmetry group and by the result of D. J. Kleitman, M. Edelberg and D. Lubell [213] one can show that

$$\binom{m}{[\frac{m}{3}]} 2^{m-[\frac{m}{3}]}$$

is the size of a maximum antichain in S^m.

DEFINITION 4.2.7. Let P be a fixed finite poset and let L_n denote a linearly ordered set of n points. Let $n_P(a, b)$ denote the number of ordered pairs (a, b) in P such that $a \le b$, in the partial order on P.

PROPOSITION 4.2.7. The number of incomparable pairs in $P \times L_n$ is at most 2 to the power

$$\binom{|P|n}{2} + n_P(a, b)\binom{n+1}{2} + |P|n$$

Proof. This follows from the fact that the number of ordered pairs $(a, b) \in P \times L_n$ such that $a \le b$ is $n_P(a, b)\binom{n+1}{2}$. □

PROPOSITION 4.2.8. The number of similarity relations on $P \times L_n$ is at most

$$\binom{|P|n}{2} - n_P(a, b)\binom{n+1}{2} + |P|n + n_P(a, b)(\log_2 C_{n+1})$$

where C_{n+1} is the (n + 1)st Catalan number.

Proof. Exercise.

LEMMA 4.2.9. Let $n_{\underline{n}}(f)$ denote the number of nondecreasing functions f from $\underline{n}$ to itself such that $x + d \geq f(x) \geq x$ for all x. Then fixed d, $n_{\underline{n}}(f) = \lambda_1^{n+o(n)}$ where λ_1 is the largest eigenvalue of the $(d + 1) \times (d + 1)$ (0, 1)-matrix

$$M = \begin{bmatrix} 1 & 1 & 0 & \cdots & 0 \\ 1 & 1 & 1 & \cdots & 0 \\ 1 & 1 & 1 & \cdots & 0 \\ & & \cdots\cdots & & \\ 1 & 1 & 1 & \cdots & 1 \end{bmatrix}$$

Proof. Let $n_{\underline{n}}(f, r)$ be the number of nondecreasing functions from $\underline{n}$ to $\underline{n + d}$ such that $x + d \geq f(x) \geq x$ for each x and $f(n) = n + r$ for $r = 0, 1, \ldots, d$. Then

$$\sum_{r=0}^{d} n_{\underline{n-d}}(f, r) \leq n_{\underline{n}}(f) \leq \sum_{r=0}^{d} n_{\underline{n}}(f, r)$$

Thus it will suffice to prove that for each r, $n_{\underline{n}}(f, r) = \lambda_1^{n+o(n)}$.

There is a recursion relation

$$n_{\underline{n+1}}(f, r) = \sum_{i=0}^{r+1} n_{\underline{n}}(f, i)$$

Thus $(n_{\underline{n+1}}(f, 0), n_{\underline{n+1}}(f, 1), \ldots, n_{\underline{n+1}}(f, d))$ can be obtained from $(n_{\underline{n}}(f, 0), n_{\underline{n}}(f, 1), \ldots, n_{\underline{n}}(f, d))$ by multiplying by the transpose of the matrix given in the lemma. Since $\lim_{k\to\infty} M^k > 0$, each of the entries in its nth power has the form $\lambda_1^{n+o(n)}$. Thus also the $n_{\underline{n}}(f, r)$ will have this form for each r. □

REMARK 4.2.3. We note that the eigenvalues of matrix mentioned in the proof of the above lemma are known to be

$$4\cos^2 \frac{2k\pi}{2d+6}$$

THEOREM 4.2.10 [206]. The number of similarity relations on $P \times L_n$ is

$$\binom{|P|n}{2} - n_P(a, b)\binom{n+1}{2} + |P|n + 2n_P(a, b)n + o(n)$$

Proof. Let S_1, S_2, S_3 respectively denote the partially ordered subsets of $(P \times L_n)' \times (P \times L_n)$ consisting of pairs (a, b, c, d) such that (i) $(a, b) < (c, d) \in P \times L_n$ for S_1, (ii) $(a, b) \le (c, d) \in P \times L_n$ for S_2, and (iii) $a \le c \in P$ and $b < d \in L_n$ for S_3. Then S_3 is a full partially ordered subset of S_1 and S_1 is a full partially ordered subset of S_2. Here S' denotes the poset obtained from S by reversing the order relation.

Proposition 4.2.8 shows that there is an upper bound of this form. To obtain a lower bound we need only show that the log number of antichains on the poset S_3 is at least $2n_P(a, b)n + o(n)$. The poset S_3 is the product of $S_q(L_n)$ and a poset of $n_P(a, b)$ elements. For simplicity, let $k = n_P(a, b)$. By refining the latter partial order to a linear order L, we obtain the result that every antichain on $S_q(L_n) \times L_k$ gives an antichain on S_3. Antichains on $S_q(L_n) \times L_k$ are in one-to-one correspondence with k-tuples $(f_1, \dots, f_k)$ of nondecreasing functions from $\underline{n}$ to itself such that $f_i(x) \ge x$ and $f_{i+1}(x) \ge f_i(x)$ for all x and all i such that these inequalities are meaningful.

For any $h > 0$, choose d large enough so that the eigenvalue λ_1 considered in Lemma 4.2.9 is larger than $4 - h$. It follows from Lemma 4.2.9 that there exists a number t such that $n > t$ the number of nondecreasing functions g from $\underline{n}$ to itself such that for all x, $x + jd \ge g(x) \ge \min\{x + (j - 1)d, n\}$ is at least $(4 - 2h)^n$, where j is any number from 1 to k. Thus the number of k-tuples $(f_1, \dots, f_k)$ such that $x + jd \ge f_j(x) \ge \min\{x + (j - 1)d, n\}$ is at least $(4 - 2h)^{kn}$. Thus for all $n > t$ the number of antichains on $S_q(L_n) \times L_n$ is $(4 - 2h)^{kn}$. Since h was arbitrary, the number of antichains on $S_q(L_n) \times L_n$ is greater than or equal to some function of the form

$$4^{kn+o(n)}$$

This proves the theorem. □

In closing this section, we shall present semiorders, which also have a connection with the Catalan numbers.

DEFINITION 4.2.8. A *semiorder* is an irreflexive and transitive binary relation R such that (i) x R y and s R t imply x R t or s R y, and (ii) x R y and y R z imply w R z or x R w for all w.

THEOREM 4.2.11 [206]. An irreflexive Boolean matrix A is the matrix of a semiorder if and only if there exists a permutation matrix P such that $P^TAP = B$ satisfies $b_{ii} = 0$, $B^2 \le B$, and $B_{i*} \le B_{j*}$ and $B_{*i} \ge B_{*j}$ for $i < j$.

Proof. The first condition applied to any matrix ensures that the rows and the columns are each linearly ordered under the usual partial order on Boolean vectors. (It is sufficient that for all A_{i*}, A_{j*} either $A_{i*} \le A_{j*}$ or $A_{j*} \le A_{i*}$. But if $A_{i*} \not\le A_{j*}$ and $A_{j*} \not\le A_{i*}$ we have s, t such that $a_{is} = 1$, $a_{js} = 0$, $a_{jt} = 1$, $a_{is} = 0$. But this violates the first condition).

Next we will show that if $A_{i*} < A_{j*}$ then $A_{*i} \ge A_{*j}$. Suppose $A_{*i} < A_{*j}$. Let s, t be such that $a_{is} = 0$, $a_{js} = 1$, $a_{ti} = 0$, $a_{tj} = 1$. Then from $a_{tj} = 1$ and $a_{js} = 1$ by the second condition either $a_{th} = 1$ or $a_{hs} = 1$ for all h. Set $h = i$ and we have a contradiction. Thus if $A_{i*} < A_{j*}$ then $A_{*i} \ge A_{*j}$. Likewise $A_{*} > A_{*j}$ then $A_{i*} \le A_{j*}$.

Define relation Q by i Q j if and only if $A_{i*} < A_{j*}$ or $A_{*i} > A_{*j}$. Then Q is irreflexive and transitive. So Q is a strict partial order. Choose a permutation P such that PAP^T lies strictly above the main diagonal, i.e., $iP\ Q\ P^Tj$ implies $i < j$. Let $B = PAP^T$. Then $i > j$ implies iP Q Pj is false which implies $A_{\pi(i)} \ge A_{\pi(j)}$ and $A_{*\pi(i)} \le A_{*\pi(j)}$ where π is such that $p_{i,\pi(i)} = 1$. This in turn implies $B_{i*} \ge B_{j*}$ and $B_{*i} \ge B_{*j}$. This proves necessity.

If B satisfies the given conditions B will be a semiorder and then so will A. This completes the proof. □

D. Scott and P. Suppes [353] proved that a binary relation R on a finite set is a semiorder if and only if there exists a real-valued function f such that x R y if and only if $f(x) \geq f(y) + 1$. Other proofs of this theorem are due to D. Scott [352] and P. Suppes and J. Zinnes [378], I. Rabinovitch [293].

THEOREM 4.2.12 [353]. A binary relation R on $\underline{n}$ is a semiorder if and only if there exists a permutation π of $\underline{n}$ and a nondecreasing function g from $\underline{n}$ to $\underline{n}$ such that $g(x) \geq x$ for each x and x R y if and only if $\pi(x) > g\pi(y)$.

Proof. Exercise.

COROLLARY 4.2.3. Isomorphism classes of semiorders are in one-to-one correspondence with nondecreasing function f from $\underline{n}$ to $\underline{n}$ such that $f(x) \geq x$ for each x.

COROLLARY 4.2.14. There exist C_n isomorphism classes of semi-orders on $\underline{n}$ where C_n is the nth Catalan number.

4.3 Connective Relations on General Partially Ordered Sets

Connectivity relations are also symmetric, reflexive binary relations associated with partial orders. But for connective relations, if x R y and one of a, b is between x, y and the other is not, then a R b is false. Thus whereas a similarity relation is "filled in," a connective relation cannot "cross itself." The following is the matrix of a connective relation:

$$\begin{bmatrix} 1 & 0 & 0 & 1 \\ 0 & 1 & 1 & 0 \\ 0 & 1 & 1 & 0 \\ 1 & 0 & 0 & 1 \end{bmatrix}$$

For a linearly ordered set, there is a one-to-one correspondence between similarity relations and connective relations. This is not true in general, but we establish results similar to those of the last section.

DEFINITION 4.3.1. Let P be a finite poset and let x P y denote the relation $x \leq y$ in the partial order. A binary relation R on P is a *connective relation* if and only if R is reflexive, symmetric and whenever x P z and z P y and y P w and $x \neq y$ and $y \neq w$ and x R y we have $z \not R w$.

If P is a linearly ordered set, D. G. Rogers [217] has shown there are one-to-one correspondence between connective relations, planar rhyming schemes, and similarity relations. The number of relations of any of the three types on a linearly ordered set of n elements is C_n, the nth Catalan number. Such relations are of interest in mathematical linguistics and have some other applications.

We first show that the one-to-one correspondence between connective relations and similarity relations does not hold true for arbitrary posets. The following results are due to K. H. Kim, D. G. Rogers and F. W. Roush [186].

PROPOSITION 4.3.1. Let P be the poset on n + m + 1 elements $x_1, \ldots, x_n, y, z_1, \ldots, z_n$ such that $a > b$ if and only if a is some x_i and b is y or some z_j, or else a is y and b is some z_j. Then the number of connective relations on P is

$$2^{t-n} - 2^{t-n-m} + 2^{t-n-m}(1 + 2^{-m})^n$$

where $t = \binom{n+m+1}{2}$.

Proof. The number t is the number of unordered pairs of elements of P. The condition of being a connective relation is equivalent to saying that if $(x_i, y) \in R$ then no pair (y, z_j) or (x_i, z_j) belongs to R where R is the same as in Definition 4.3.1.

The number of connective relations with no pair of the form (x_i, y) is 2^{t-n}. The number of connective relations with exactly s pairs of this form is

$$\binom{n}{s}2^{t-n-m-sm}$$

A summation proves the proposition. □

We remark that this formula is not symmetric in n, m, yet the corresponding formula for similarity relations would be symmetric.

In the last section we estimated the numbers of similarity relations on certain posets. Firstly, we shall show that such estimates hold for connective relations on these posets.

DEFINITION 4.3.2. If D is a binary relation, a set S is an *independent set* for D if and only if x D y does not hold for any x, y ε S.

DEFINITION 4.3.3. The binary relation D on $S_q(P)$ is given by (x, y) D (a, b) if and only if one of the following hold: (i) x P a, a P y, y P b and y ≠ b, (ii) y P a, a P x, x P b and x ≠ b, (iii) x P b, b P y, y P a and y ≠ a, and (iv) y P b, b P x, x P a and x ≠ a.

THEOREM 4.3.2 [186]. Let S be the set of unordered pairs (x, y) ε $S_q(P)$ such that (x, y) ε R. Then S is a D-independent set. Conversely, let S be any D-independent set. Then the binary relation R given by (x, y) ε R if and only if x = y or (x, y) ε S is a connective relation.

Proof. The proof follows from Definitions 4.3.1 and 4.3.2. □

COROLLARY 4.3.3. Let S_1 be the set of incomparable pairs in a poset P. Let S_2 be the set of comparable pairs in P. Then the number of connective relations on S_1 is

$$2^{|S_1|}$$

times the number of D-independent subsets of S_2.

Proof. Adding incomparable pairs to any subset of S will not alter its being or not being an independent set. □

PROPOSITION 4.3.4. The number of connective relations on V_m is at least

$$2^{2^{2m-1}} - 3^m + 2^{m-1} + c_m - 1$$

where c_m is the coefficient of x^m in $(1 + x + x^2)^m$.

Proof. Let K be the set of all ordered pairs $(x, y) \in V_m$ such that $x \leq y$. The set H of comparable pairs in $S_q(V_m)$ can be regarded as the subset of K consisting of pairs (x, y) such that $x < y$. We can also regard K as the mth power of the set (0, 0), (0, 1), (1, 1). The subset of K of elements with i components (0, 0) and j components (0, 1) and k components (1, 1) for all i, j, k such that $j > 0$ and $i = k$, is D-independent and contained in the image of H. The cardinality of this set is $c_m - 1$ or c_m. □

We next consider the products with a linear order. Our results in this case are less complete than for similarity relations in this case.

THEOREM 4.3.5 [186]. The number of connective relation on $P \times L_n$ is divisible by

$$2^{\binom{|P|n}{2} - n_P(a, b)\binom{n+1}{2} + |P|n}$$

and is at most

$$2^{\binom{|P|n}{2} - n_P(a, b)\binom{n+1}{2} + |P|n + n_P(a, b)(\log_2 C_{n+1})}$$

where C_{n+1} is the (n + 1)st Catalan number.

Proof. The first statement can be established by counting the number of incomparable pairs in $P \times L_n$.

Let T_1 denote the set of comparable pairs in $P \times L_n$. Let T_2

denote the number of ordered pairs (a, b) where a, b ε $P \times L_n$ such that $a \leq b$.

To prove the second statement of the theorem, it will suffice to show that the number of D-independent sets of T_1 is at most

$$C_{n+1}^{n_P(a, b)}$$

The relation D on T_1 extends to a relation on T_2 given by (a, b) D (c, d) if and only if $a \leq c \leq b < d$. Let Q be the relation on T_2 given by $((p_1, n_1), (p_2, n_2))$ Q $((p_3, n_3), (p_4, n_4))$ if and only if $p_1 = p_3$, $p_2 = p_4$, $n_1 \leq n_3 \leq n_2 < n_4$ where $p_i \varepsilon P$ and $n_i \varepsilon L_n$. Then every D-independent set will be Q-independent.

Under Q, T_2 splits as a disjoint union of $n_P(a, b)$ copies of a set M consisting of pairs (n_1, n_2) such that $n_1 \leq n_2$ on which Q is defined by (n_1, n_2) Q (n_3, n_4) if and only if $n_1 \leq n_3 \leq n_2 < n_4$. If we send all pairs (n_1, n_2) to $(n_1, n_2 + 1)$ we obtain a D-independent set in $S_q(L_{n+1})$. By D. G. Rogers [318] the number of D-independent sets in $S_q(L_{n+1}) = C_{n+1}$. This proves the theorem. □

It appears that in order to establish a good lower bound it will be necessary to consider a disjoint union of two sets $S_q(L_{n+1})$ or M on which D is not a disjoint union of its restrictions to the two parts. To be precise, we consider the set S_T of triples (a, b, c) such that $1 \leq a \leq b \leq n$ and $c = 0$ or $c = 1$. We define a relation D in S_T by (a, b, c) D (a', b', c') if and only if $a \leq a' \leq b \leq b'$ and $c \leq c'$. Then it is desired to compute the number of D-independent pairs in S_T. This number is less than or equal $(C_n)^2$ since if we required $c = c'$ in the definition of D it would be exactly $(C_n)^2$.

In order to study S_T we will use some geometric language appropriate to planar rhyming schemes. A connective relation R on $\underline{n}$ can be represented by drawing an arc between any two R related numbers. such arcs cannot intersect and cannot meet except at a number which is the right end of both. For instance

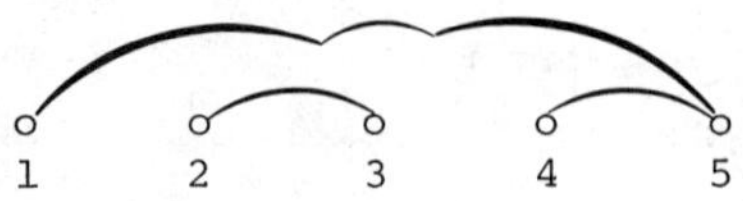

represents a connective relation. We say that a number i is open if one could add an element 0 at the left of 1, 2, 3, 4, 5 and draw an arc from 0 to i, and the result would be a connective relation. There is always at least one open number, the largest in the chain.

For S_T we have a pair of such diagrams. For instance, the diagram

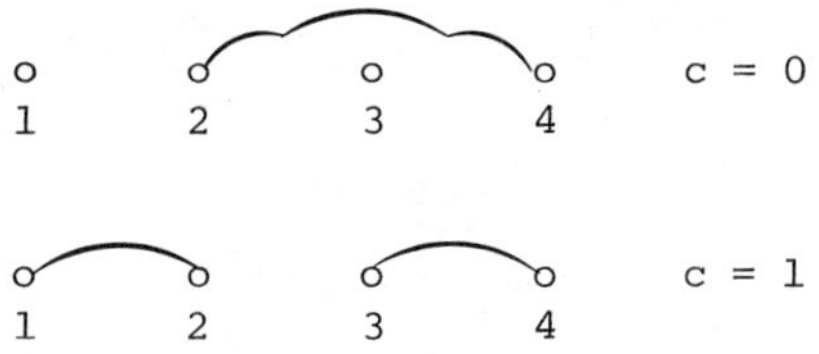

is allowed in S_T, though if the c = 0 and c = 1 diagram were interchanged, the result would not be a connective relation.

We next consider how many ways such a diagram might be extended by adding two zeros. The lower 0 can be connected to any i which is open for that diagram (we call such i, *s-open*). The upper 0 can be connected to any j such that $j \geq i$ and j is open in both diagrams (we say *t-open*).

To every pair of diagrams we associated a word in t and s. This word gives the sequence of t and s-open numbers. For instance for the diagram above we have the word st since the number 2 is s-open and the number 4 is t-open. The actual values of t and s-open numbers are disregarded.

We next ask, if a point 0 is added to extend a pair of diagram, what extensions of the associated words are possible. If no arrows are drawn from 0, the effect is to add t to the word. If only an arrow from zero occurs in the upper diagram, the t is converted to s and all to its left in the word are converted to s. If a lower arrow is drawn from 0 to i and an upper arrow from 0 to j, $j \geq i$, all letters in the word which occur to the left of i are dropped and all letters which occur to the left of j are changed to s. Thus, we proved the following result.

PROPOSITION 4.3.6. Start with the word t. Go through a series of n - 1 steps each time adding a t, then dropping all letters to the right of some letter, then converting all letters to the right of some t to s. Count the number of words so obtained, counting a word k times if it has been obtained k ways. This number is the number of D-independent subsets of S_T.

This procedure can also be represented in matrix terms. We index the rows and columns of an infinite-dimensional matrix on words in t, s whose right-most letter is t. We enter a 1 in location (i, j) if and only if j can be obtained from i by adding t, dropping all letters, then changing all letters to the right of some letter to s. Let W be the resulting matrix. Let v be the vector which has a 1 in location t and zeros elsewhere. Then the weight of vW^{n-1} equals the number of D-independent subsets of S_T.

If we could obtain the limit of the largest eigenvalues of finite submatrices of W, this might help to replace Lemma 4.2.9. Or if we could find an eigenvalue of W itself for which W has a nonnegative eigenvector.

Before closing this section, we should point out that Ju. A. Schreider [340] studied mathematical linguistics from the viewpoint of the various order relations.

4.4 Applications to Economics

Binary relations are important in economics as the preference relations of individuals or groups. The behavior of a market is determined by the resources and preferences of the individual participants. Moreover the fairness of an allocation of goods is perhaps measured by the preferences of the individuals. These ideas also reach out to game theory, where players' preferences for different payoffs are significant, and to political science, where voting methods must in some manner reflect the preferences of the voters.

Preference relations of individuals in economics are most frequently represented by *weak orders*, relations which are reflexive, transitive, and complete. These are quasi-orders whose equivalence classes are linearly ordered.

Group preferences are usually required to be transitive and reflexive, and sometimes also complete. In addition, they should be *Pareto* (if everyone prefers x to y so does the group), nondictatorial, and perhaps the group choice between two alternatives x, y should depend only on the individual preferences between these two. Kenneth Arrow proved that no such group preference relation exists for more than two alternatives, if all combinations of individual preferences are allowed.

A. Gibbard and M. Satterthwaite proved in addition that every nondictatorial social choice function having at least three different alternatives as values is manipulable. This means that for some combination of individual preferences, some individual would gain an alternative which is more desirable to him as group choice, if he misrepresented his actual preferences.

Here we show that if we allow a choice of more than one alternative, and define preferences among subsets in a certain way this need not be so. J. Kelly has done similar work. However A. Gibbard has also extended his theorem to lotteries, which means that in reality manipulability may be the rule for sets of more than one element as well as choices of a single alternative.

DEFINITION 4.4.1. Let X be a set containing n elements. A *preference relation* is a binary relation R on X such that (i) $(x, x) \in R$ for all x, (ii) if $(x, y) \in R$ and $(y, z) \in R$ then $(x, z) \in R$, and (iii) for all (x, y) either $(x, y) \in R$ or $(y, x) \in R$.

Such a relation can be interpreted as follows: $(x, y) \in R$ means that the individual considers x to be at least as good as y, where $x, y \in X$. Thus, a linear order is a preference relation R which in addition satisfies antisymmetricity. That is, a linear order on X can be regarded simply as a ranking of the elements of X as first, second, third, ... , nth.

DEFINITION 4.4.2. Let E be an equivalence relation on X and let R_E be a relation on the set of equivalence classes of E. The *inflation* of R_E under E is the relation R on X such that $(x, y) \in R$ if and only if $(\bar{x}, \bar{y}) \in R_E$ where $\bar{x}$ denotes the equivalence class of x. The relation E is sometimes called the *indifference relation* of R, and the relation P on X such that $(x, y) \in P$ if and only if $(x, y) \in R$ but $(x, y) \notin E$ is called *strict preference*.

The following propositions which establish a relationship between Boolean matrices and preference relations are due to K. H. Kim and F. W. Roush [206].

PROPOSITION 4.4.1. A Boolean matrix A is the matrix of a preference relation if and only if (i) $A^2 = A$, and (ii) $A + A^T = J$.

Proof. Computation. □

PROPOSITION 4.4.2. The strict preference relation P on X has matrix given by $B = (A^T)^C$ where A is the matrix of the preference relation R. Conversely $A = (B^T)^C$. A matrix B is the matrix of a strict preference relation if and only if (i) $B \odot B^T = 0$, and (ii) $(B^C)^2 = B^C$. Here $\odot$ denotes the elementwise product.

Proof. Let M denote the matrix of equivalence relation E. Then $B \odot A^T = B \odot A \odot A^T = B \odot M = 0$. Thus $B \le (A^T)^C$. Furthermore $B + A^T = B + (M + B)^T = B + M + B^T = B + M + M^T + B^T = (B + M) + (M + B)^T = A + A^T = J$. Hence $B = (A^T)^C$. This implies $A = (B^T)^C$. Thus, using these expressions (i) and (ii) of the present proposition are equivalent to (i) and (ii) of Proposition 4.4.1. □

The following proposition is a direct consequence of Definition 4.4.2.

PROPOSITION 4.4.3. Let R be a preference relation on X. Let E be the binary relation such that $(x, y) \in E$ if and only if both $(x, y) \in R$ and $(y, x) \in R$. Then E is an equivalence relation and R is the inflation of a linear order under E. Conversely every inflation of a linear order is a preference relation.

DEFINITION 4.4.3. Let P_X denote the set of preference relations on X. Let P_X^k denote the set of all ordered k-tuples of preference relations on X. Then a *social welfare function* (*voting procedure*, *aggregation method*) is a function from P_X^k to P_X for some $k \in \underline{n}$.

Thus if f is a social welfare function, for any set of k individuals preferences $p_1, \ldots, p_k$ of these individuals, f will give the preferences of the group.

In terms of Boolean matrix theory, let $\mathcal{A}_n$ denote the set of Boolean matrices $A \in B_n$ such that (i) $A \odot A^T = 0$, (ii) $(A^C)^2 = A^C$. Let $\mathcal{A}_n^k$ denote the set of all ordered k-tuples of $\mathcal{A}_n$. Then a social welfare function is equivalent to a function from $\mathcal{A}_n^k$ to $\mathcal{A}_n$. We will denote a typical element of $\mathcal{A}_n^k$ by $(A(1), \ldots, A(k))$.

In the following we shall present the famous Arrow's Impossibility Theorem in terms of Boolean matrices due to K. H. Kim and F. W. Roush [206].

THEOREM 4.4.4 [12]. For $n \geq 3$ there is no social welfare function f satisfying the following conditions on all n-tuples of linear order matrices.

(1) (PARETO PRINCIPLE). The matrix $f(A(1), \ldots, A(k))$ is greater than or equal to the elementwise product of $A(1), \ldots, A(k)$, and less than or equal to their elementwise sum.

(2) (INDEPENDENCE OF IRRELEVANT ALTERNATIVES). Suppose for some i, j, $A(m)_{ij} = A'(m)_{ij}$ for $m = 1, \ldots, k$. Then $f(A(1), \ldots, A(k))$ and $f(A'(1), \ldots, A'(k))$ have the same (i, j)-entry.

(3) (NONDICTATORIAL). For all m, the function f is not the function $(A(1), \ldots, A(k)) \to A(m)$.

For a proof see R. D. Luce and H. Raiffa [237]. Two apparent ways in which something like a social welfare function might be found are (i) to weaken the conditions above and (ii) to require that f be defined only on a subset of A_n^k. D. Black [24], K. Inada [161], B. Ward [396], A. K. Sen [357], V. J. Bowman and C. S. Colantoni [38] have proposed suitable subsets of A_n^n. For instance Black proved that majority decision, for an odd number of voters, is a social welfare function on the subset of single-peaked preferences.

DEFINITION 4.4.4. A preference relation on $\underline{n}$ is *single-peaked* if and only if for any i, j, k ε $\underline{n}$ such that $i < j < k$, either all three alternatives are indifferent or j is strictly preferred to one of the others.

Geometrically, suppose one arranges the alternatives according to magnitude and graphs ranks that the order relation assign to the alternatives. Then the resulting curve will increase to a maximum and then decrease.

For three alternatives M. B. Garman and M. I. Kamien [122] proved the conjecture of G.-Th. Guilbaud [132] that as n tends to infinity the probability of not having a majority winner approaches 0.0877... .

It has seemed to some that while the first and the third conditions of Theorem 4.4.4 are indispensable, the necessity of the second condition is not so apparent. For instance in the 3 individual, 3 alternative case where individual a ranks the alternatives 123, individual b ranks the alternatives 231, and individual c ranks the alternatives 321, by symmetry the group preference matrix should be

$$\begin{bmatrix} 1 & 1 & 1 \\ 1 & 1 & 1 \\ 1 & 1 & 1 \end{bmatrix}$$

If independence of irrelevant alternatives holds the result on 1, 2 will be the same as for a: 123; b: 213; c: 123. Yet in this latter case it is natural to regard 1 as being preferred by the group. A replacement that has proved satisfactory is the assumption that no individual can manipulate the decision, i.e., gain a more favorable social decision by misrepresenting his own preferences. Of course this happens in democracies: a voter with extreme views may vote as if his real favorite was a more popular candidate leaning somewhat in his direction. A number of authors have recently proved impossibility theorems about non-manipulable social choice functions, for instance (S. Barbera [16]).

Another condition that might perhaps not be necessary is that the image of f be transitive. To say that the group prefers x to y could mean that if faced with x and y only, the group would choose x. Yet the other alternatives, as above, may be relevant.

Thus in some ways it is more natural to consider instead of a social welfare function, the following.

DEFINITION 4.4.5. A *choice function* on a set of alternative X is a mapping g from subsets of X to subsets of X such that for all $S \subset X$ we have $g(S) \subset S$.

DEFINITION 4.4.6. Let S_X^k denote the set of k-tuples of antisymmetric preference relations on X. A *social choice function* for k individuals and a set X of n alternatives is a mapping from S_X^k to the set of choice functions on X. (We assume antisymmetry only in order to prove Theorem 4.4.6. Otherwise, there is no theoretical reason for assuming it).

The social choice function represents the g(S) of alternatives which the group considers best if it is confronted with a subset S of the alternatives. The group is considered indifferent to which alternatives among g(S) it would prefer (they are tied for first place). In practice it is usually desired to assume that g(S) is nonempty whenever S is nonempty. Frequently g is required to be defined only on a subset of the set of all subsets of X.

DEFINITION 4.4.7. Let R be a preference relation on X. A function h from X to $\mathcal{R}$ is called a *nondecreasing system of weights* on X if and only if $h(x) \geq h(y)$ whenever $(x, y) \in R$. Let $S, T \subset X$. Then S is **-preferred* to T if and only if for every nondecreasing system of weights h(x) the average weight of an element of S is larger than the average weight of an element of T.

DEFINITION 4.4.8. For a set X of alternatives and a positive integer m, let mX denote a set of order $m|X|$ which we regard as containing m copies of each alternative of X. For a preference relation R on X, define a preference relation on mX by giving a copy of an alternative the same rank as that alternative. For a subset S of X and an integer $0 < r \leq m$ let rS denote any subset of X containing exactly r copies of each alternative in S.

PROPOSITION 4.4.5. The relation of *-preference is reflexive and transitive. For any r, s > 0, S is *-preferred to T if and only rS is *-preferred to sT. Suppose $|S| = |T|$. Then S is *-preferred to T if and only if there exists a one-to-one function ψ from S onto T such that $(x, \psi(x)) \in R$ for each R.

Proof. The first assertion follows from the definition of *-preference. The assertion about rS and sT likewise follows from the definition. Suppose $|S| = |T|$ and S is *-preferred to T. Arrange the elements of S and of T an order such that if x is strictly preferred to y then x precedes y. Let ψ be the function from S to T sending the ith element of S to the ith element of T for each i. Suppose $(x, \psi(x)) \notin R$ for some x. Assign to all elements $z \in X$ such that $(z, \psi(x)) \in R$ the weight 1 and to all other elements of X the weight 0. Then the average weight of S. This is a contradiction. So if S is *-preferred to T a function ψ exists. It follows from the definition that if a ψ exists, then S is *-preferred to T. □

DEFINITION 4.4.9. A social choice function is *nondictatorial* if it is not a function of the choices of some single individual.

DEFINITION 4.4.10. A subset S of X is *strictly *-preferred* to T of X is S is *-preferred to T but T is not *-preferred to S.

DEFINITION 4.4.11. A social choice function satisfies the *Pareto condition* if and only if whenever $S \subset Y$ and there exists $T \subset Y$ such that every individual in the group *-prefers T to S and some individual strictly *-prefers T to S then $g(Y) \neq S$ where $Y \subset X$.

DEFINITION 4.4.12. A social choice function cannot be *manipulated* if there does not exist an individual i and a subset Y of X such that if all other individuals are honest and i misrepresents his preferences in some way, the resulting set g(Y) will be strictly *-preferable for i to the set which would result if everybody were honest.

DEFINITION 4.4.13. A social choice function is *symmetric in the individuals* if it is invariant under replacement of $(A(1), \ldots, A(k))$ by $(A(\pi(1)), \ldots, A(\pi(k)))$ for any permutation π.

DEFINITION 4.4.14. A social choice function is *symmetric in the alternatives* if and only if $g(P^T A(1)P, \ldots, P^T A(k)P) = \pi(g(A(1), \ldots, A(k)))$ for any permutation matrix P, where π is the permutation such that $p_{i\pi(i)} = 1$ for i = 1 to n.

THEOREM 4.4.6 [206]. Let $Y \subset X$. For any set of at least two individuals there exists a social choice function which is Pareto, nondictatorial, nonmanipulable, nonempty for $Y \neq \emptyset$, symmetric in both the individuals and the alternatives. Specifically such a function is given by taking g(Y) to be all alternatives $a \in Y$ such that some individual does not rank any other alternative in Y strictly above a.

Proof. It can be checked that this social choice function is nondictatorial, nonempty for $Y \neq \emptyset$, and symmetric in the individuals and the alternatives. Next we will deal with the proof that it is Pareto. Suppose on the contrary that there existed a set of preferences $(A(1), \ldots, A(k))$ and a subset Y of X and a subset S of Y such that every individual *-prefers S to g(Y). Suppose x is the alternative which ranks first one some individual's list, i.e., $x \in g(Y)$. Then if that individual *-prefers S to Y, x must be in S. Thus $g(Y) \subset S$. By Proposition 4.4.5, $|S|g(Y)$ will be *-preferred by every individual to $|g(Y)|S$. However if $|S| > (Y)$ then $|g(Y)|S$ will contain fewer copies of x than $|S|g(Y)$, and the individual cannot *-prefer $|g(Y)|S$ to $|S|g(Y)$. Thus $|S| = |g(Y)|$. Thus $S = g(Y)$, so that S will not be strictly *-preferred by anyone to g(Y). This proves the Pareto condition.

Then we must prove that the given social choice function cannot be manipulated. Suppose individual i by misrepresenting his preferences could obtain a strictly *-preferable choice set. Then he must have changed his nomination for first choice among the elements of Y: otherwise g(Y) cannot be changed. But if he does this then he cannot obtain a *-preferable set. This proves the theorem. □

The following are some other social choice functions which satisfy the theorem.

(1) Let g(Y) be as in Theorem 4.4.6 unless a majority of the individuals prefer a certain alternative in Y to all other alternatives in Y. In that case choose the majority alternative.

(2) Let g(Y) be as in Theorem 4.4.6 unless there exist two alternatives each of which is preferred to all other alternatives in Y by more than $\frac{1}{3}$ of the voters. In that case choose the set consisting of the two alternatives.

(3) Let g(Y) be the set of alternatives in Y which are not least in anyone's list among the alternatives in Y, unless that set is empty. If the set is empty take $g(Y) = Y$.

The extension to the case where the preferences of individuals are not antisymmetric, i.e., where individuals are allowed to say that they like certain alternatives equally well, appears to be quite difficult. For instance if we have two individuals a, b and three alternatives 1, 2, 3 and the rankings are

a	b
12	13
3	2

then g({1, 2, 3}) = {1, 2, 3} does not satisfy the Pareto condition, because g({1, 2, 3}) = {1} is strictly *-preferable for both individuals.

A different definition of *-preferable could be adopted. That is one might assume each individual has his own system of weights and he prefers certain sets above others in exact accordance with their average weight. This case was dealt with by a theorem of A. Gibbard on the lotteries.

B. G. Mirkin [256] has also used Boolean matrices in the study of social choice, in particular to illustrate various order relations on a 3-element set.

For further details see K. J. Arrow [12], D. Black [24], P. C. Fishburn [113], A. Jaeger [162], R. D. Luce and H. Raiffa [237], W. H. Riker and P. C. Ordeshook [314], J. Rothenberg [325], A. Wieczorek [400], and A. K. Sen [356, 357].

There is also a considerable literature dealing with characterization of different types of choice functions.

4.5 Hall Relations

In closing this chapter we shall briefly mention another important kind of order relations known as Hall relations. Hall relations are of great importance in combinatorics, operations research, and Boolean matrix theory.

DEFINITION 4.5.1. Let $X = \{x_1, \ldots, x_n\}$. Let $X_1, \ldots, X_n$ be subsets of X. A *system of distinct representatives* (*SDR*) is a sequence $s_1, \ldots, s_n$ such that $s_i \in X_i$ for each i and for $i \neq j$, $s_i \neq s_j$.

DEFINITION 4.5.2. A binary relation H on X is a *Hall relation* if and only if the sets $x_1H, \ldots, x_nH$ have an SDR.

For general reference on SDR's, see H. J. Ryser [331], and C. Berge [21].

A Boolean matrix A is a Hall matrix if and only if it is the matrix of a Hall relation, if and only if $A \geq P$ for some permutation matrix P. (See Definition 2.4.2). It is immediate that the Hall matrices form a semigroup. K. H. Kim [180] characterized L, R, H, D, J-classes in this semigroup, found their cardinalities, and proved idempotent Hall matrices are in one-to-one correspondence with quasi-orders.

K. H. Kim and F. W. Roush [206] related SDR's to the structure of semigroups. Consider a Rees matrix A semigroup with 0 (see A. H. Clifford and G. B. Preston [56]), with group G = {1}. A Boolean matrix B interpreted as a set of H-classes, represents a subsemigroup if and only if $BAB \leq B$. They proved this result.

THEOREM 4.5.1. Suppose that, when the columns of A are regarded as subsets of $\underline{n}$, A has an SDR, and that no row or column of A contains less than two ones. Then for B satisfying $BAB \leq B$ the matrix B must contain nm ones or have at most $nm - n$ ones. Here m is the number of columns of A.

Exercises

1. Compute topologies on a set of two points.
2. Compute T_0-topologies on a set of three points.
3. Compute topologies on a set of three points.
4. How many reflexive matrices are there in B_n ?
5. Prove that a relation is transitive if and only if its matrix satisfies $A^2 \leq A$.

6. Give an example of a 5×5 matrix with at least two ones in each row and each column, which is not a Hall matrix.
7. Give examples of 3×3 matrices which are similarity relations and which are connective relations on the poset $1 < 2 < 3$. Draw their graphs.
8. Derive good upper and lower bounds for Stirling's number of the second kind.
9. Find an asymptotic formula for t_n, where t_n is the number of transitive relations on $\underline{n}$.
10. Prove Proposition 4.2.8.
11. Show that the eigenvalues of the matrix in Lemma 4.2.9 are

$$4\cos^2 \frac{2k\pi}{2d+6},\ 0$$

12. Count the number of Hall matrices in B_n having $n + r$ ones for $r = 0, 1, 2, 3$.
13. Let $A \in B_n$. Prove that if A is a Hall matrix it cannot have an $r \times s$ rectangle of zeros where $r + s = n + 1$. This condition is also sufficient by the Hall-König theorem.
14. Let $A \in B_n$. Prove that A is a Hall matrix if and only if for all vectors v, vA has at least as many ones as v.
15. Let $A \in B_n$. Prove that if B is a Hall matrix, AB and BA have at least as many ones as A.
16. Prove an idempotent Hall matrix is reflexive.
17. What is the simplest partial order which is not a semiorder ?
18. Prove that a non-Hall matrix has Schein rank less than n.
19. Prove the Hall-König theorem using transcendence basis for a field, as follows. We must show that if M is not greater than or equal to any permutation matrix then M has an $r \times s$ rectangle of zeros, where $r + s = n + 1$. Let A be a matrix over the real numbers obtained from M by replacing all the 1 entries of M by algebraically independent transcendental real numbers. Then Det (A) = 0. Thus there exists a nonzero vector v such that vA = 0, and some row is a linear combination of other rows. Show that this expresses the entries of A in terms of a smaller number

algebraically independent quantities, unless A has a rectangle of zeros with $r + s = n + 1$.

20. What is the number of SDR's in matrix terms ?
21. Write a formula for the exact number of elements in the D-class of partial permutations of rank k.

CHAPTER FIVE

ASYMPTOTIC FORMS

We present the asymptotic forms of powers of Boolean matrices. By the "asymptotic forms" of the powers of a given matrix we mean essentially the complete description of the location of zeros in the asymptotic development of the powers.

The present characterization of the asymptotic forms of finite dimensional matrices is suggested by relation-theoretic considerations [399]. We begin with a graph-theoretic characterization of the asymptotic forms of powers of matrices.

In this chapter we look at the series of powers of a matrix in several different ways. Some overlap is unavoidable when we treat the same subject by different methods. A summary of results about asymptotic forms of matrices is provided in Theorem 5.4.25. We conclude with several applications of this theory: small group structure in sociology, clustering, diffusion of information, and communications nets, design of combinatorial relay circuits, and finite-state nondeterministic automata.

5.1 Graphical Characterization

In this section we are concerned with using matrices to obtain information about connectedness properties of graphs. Recall (Def. 1.2.6) that a directed graph (digraph) is really the same as a Boolean matrix. In this chapter we assume all graphs are digraphs.

The results in this section are contained in F. Harary, R. Z. Norman and D. Cartwright [140].

DEFINITION 5.1.1. The *reachability matrix* of a graph is the matrix R such that $r_{ij} = 1$ if v_j is reachable from v_i by some sequence of directed edges (every point is considered reachable from itself).

DEFINITION 5.1.2. A *path* in a graph is a sequence of distinct vertices $v_0, v_1, \ldots, v_n$ such that for $i = 0, 1, \ldots, n - 1$ there is a directed edge in the graph from v_i to v_{i+1}.

PROPOSITION 5.1.1. Given the adjacency matrix A, the reachability matrix R may be computed by the formula $R = (I + A)^{p-1}$ where p is the number of vertices in the graph and the matrix operations are Boolean.

Proof. $(I + A)^{p-1} = I + A + \ldots + A^{p-1}$ in Boolean matrix algebra. And j is reachable from i by a sequence of directed edges of length k if and only if the (i, j)-entry of A^k is 1, since this entry is 1 if and only if there is a sequence $i = i_1, \ldots, i_{k+1}$ such that $a_{i_1 i_2} a_{i_2 i_3} \ldots a_{i_k i_{k+1}} = 1$. □

REMARK 5.1.1. Note that there are also k-reachability matrices R_k whose (i, j)-entries are 1 if and only if there is a directed path from i to j of length less than or equal to k. So that $R_k = (I + A)^k$.

DEFINITION 5.1.3. A graph is *strongly connected* if and only if for any two vertices i, j, j is reachable from i. A graph is *weakly connected* if, disregarding the direction of the edges, there is a path in the graph from any vertex to any other vertex. A graph is *unilaterally connected* if and only if for any two of its vertices, one is reachable from the other. (In terms of Boolean matrices, a graph is unilaterally connected if and only if $R + R^T = J$ where R is the reachability matrix).

DEFINITION 5.1.4. The *strong components* of a graph G are those full subgraphs of G that are strongly connected and are not properly contained in any other strongly connected subgraph of G.

PROPOSITION 5.1.2. For any i the strong component containing v_i consists of the vertices v_j such that $(R \odot R^T)_{ij} = 1$ where $\odot$ is the elementwise product.

Proof. Exercise.

DEFINITION 5.1.5. The *condensation* C* of a graph C is a graph whose vertices are in one-to-one correspondence with the strongly connected components of C, such that there is an edge from v_i to v_j if and only if in C there is a directed edge from some point of the ith strong component to some point of the jth strong component.

If the vertices are ordered in such a way that vertices in the same strong component are next to each other, the adjacency matrix of C* can be found from that of C by partitioning it according to the strong components and replacing each block containing at least one 1 with a 1 and each identically zero block with a zero. (Except that the main diagonal entries for C* will be assumed to 0).

DEFINITION 5.1.6. A *point basis* of a graph is a minimal set of points from which every point can be reached. A *point contrabasis* is a minimal set of points such that from any point of the graph, at least one of them can be reached. A *source* is a point from which every point of the graph can be reached. A *sink* is a point which is reachable from every point of the graph.

THEOREM 5.1.3 [140]. A point basis of C consists of one point from each of the strong components which form a point basis for C*. There is a unique point basis of C*. A point of C* is in this basis if and only if in the matrix C* its column sum is 0.

Proof. The first statement follows from Definitions 5.1.5, and 5.1.5. The relation of reachability makes C* into a poset, and a point belongs in the point basis of C* if and only if it is a maximal element of this poset. These points are those into which no directed edge in the graph of C*. This proves the theorem. □

DEFINITION 5.1.7. A set V of vertices of a graph is a *fundamental set* if and only if there is some vertex v from which every point of V is reachable, and is no vertex v' from which is a set S properly containing V can be reached. A point from which all the points of a fundamental set is reachable is called an *origin*. The dual concepts are *contrafundamental set* and *terminus*.

THEOREM 5.1.4 [140]. A set is a point basis if and only if it consists of one origin from each fundamental set.

Proof. We can again consider the condensation C*. A point in C* will be an origin if and only if it is a maximal point of the poset if and only if it is in a point basis. □

THEOREM 5.1.5 [140]. A vertex v_i is an origin of some fundamental set if and only if the number of 1's in the ith column of reachability matrix R is equal to the number of 1's in the ith column of $R \odot R^T$. If it is an origin its fundamental set is the set of all points whose entry in the ith row of R is 1.

Proof. The first condition means v_i is reachable only from other points in its strong component. This means the strong component will be a maximal point of C*. The fundamental set will be all points reachable from v_i, which is equivalent to the condition of the second statement. □

REMARK 5.1.2. Note that a graph is strongly connected if and only if its reachability matrix is the universal matrix.

PROPOSITION 5.1.6 [140]. The weak component containing a vertex v_i of graph G is given by the set of 1 entries in the ith row of $R_W = (I + A + A^T)^{p-1}$, where A is an adjacency matrix of G.

An overall specification of the connectedness of a graph is given by the connectedness matrix C such that (i) $c_{ij} = 0$ if and only if there is no path from v_i to v_j even ignoring the direction of edges, (ii) $c_{ij} = 1$ if and only if there is a path from v_i to v_j if directions of edges are ignored but there is no directed path from either vertex to other, (iii) $c_{ij} = 2$ if and only if there is a directed path from one of v_i, v_j to the other, but not conversely, (iv) $c_{ij} = 3$ if and only if there is a directed path from v_i to v_j and a directed path from v_j to v_i.

It is determined by $c_{ij} = r_{ij} + r_{ji} + 1$ (addition of integers) if v_i and v_j are in the same weak component and $c_{ij} = 0$ otherwise. Here r_{ij} is an element of reachability matrix R.

DEFINITION 5.1.8. The *distance matrix* D of a directed graph is the matrix whose entries d_{ij} are the lengths of the shortest directed paths from v_i to v_j.

One way to determine the distance matrix is that $d_{ii} = 0$, $d_{ij} = \infty$ if and only if $r_{ij} = 0$ and otherwise d_{ij} is the least number s such that the sth power of adjacency matrix A has a 1 for its (i, j)-entry.

An alternative formula is $D = kR_\infty - R_0 - R_1 - \cdots - R_{k-1}$ where k is the least integer such that $R_k = R$, R_∞ is the matrix in which the 0's in R are replaced by ∞, and the operations are of integers. There is also a non-Boolean method related to dynamic programming.

From D the shortest paths from v_i to v_j can be determined. A vertex v_k lies on such a path if and only if $d_{ik} + d_{kj} = d_{ij}$.

In certain types of applications of Boolean matrices, an algorithm due to K. R. Parthasarathy [275] may be useful.

THEOREM 5.1.7 [206]. For a graph G let M_G^n be the matrix whose (i, j)-entry is the number of paths of length n from v_i to v_j. Let A be the adjacency matrix of G and let G_j be the subgraph whose adjacency matrix is G with the jth row and jth column replaced by zero row and zero column. The jth column of M_G^n equals the jth column of the matrix product $(M_{G_j}^{n-1})A$.

Proof. Equality of the jth entries in these columns: the (j, j)-entry of M_G^n is zero, since the vertices in the path cannot be distinct. But since the jth row of $M_{G_j}^{n-1}$ is zero, the jth entry in the jth column of the product will also be zero.

Equality of the ith entries for $i \neq j$: Let $V = v_i \ldots v_t$ be a path of length n - 1 and suppose there is an edge $v_t v_j$ in G. Then $v_i \ldots v_t v_j$ is a path if and only if v_j does not occur in V, i.e., if and only if V is a path in G_j. Conversely every path of length n in G can be represented in this way. The matrix interpretation of this representation is the equality in question. □

5.2 Convergent and Oscillatory Matrices

In this section we present the various properties of asymptotic forms of powers of matrices. We shall begin with a graph-theoretic characterization of the asymptotic forms of powers of matrices due to D. Rosenblatt [321].

First, the asymptotic properties of indecomposable matrices are considered. A matrix is indecomposable if and only if its graph is strongly connected. The period of indecomposable matrix is the greatest common divisor of the lengths of all cycles in its graph. Moreover an indecomposable matrix of period greater than 1 can be put into a special block-permutation form (Theorem 5.2.9). If the period is 1 the matrix converges to J (Proposition 5.2.15). If the index is 1 the block-permutation form has each block consisting entirely of ones or entirely of zeros (Theorem 5.2.18). A decomposable matrix of period 1 which is not nilpotent converges to an idempotent E such that 0 < E < J (Proposition 5.2.13). Some sufficient conditions for a matrix to have period greater than 1 are given (Theorems 5.2.11 and 5.2.12). The existence of a 1 in each row (row nonempty) carries over to powers of a matrix and is important in these proofs. A matrix is nilpotent if and only if its graph has no cycles (Theorem 5.2.10) if and only if it is conjugate to a subtriangular matrix (Proposition 7.2.14).

In the last part of this section primitive circulants and primitive companion matrices are discussed.

DEFINITION 5.2.1 [321]. A matrix $A \in B_n$ $(n \geq 2)$ is said to be *convergent* (in its powers) if, and only if, there exists in the sequence $\{A^k: k = 1, 2, \ldots\}$ a power of A^m of A such that $A^m = A^{m+1}$. It is said that A *converges* to the matrix A^m. A matrix A is said to be *oscillatory* (*periodic*) (in its powers) if, and only if, there exists in the sequence $\{A^k: k = 1, 2, \ldots\}$ a power A^m of A such that $A^m = A^{m+p}$ where p is the smallest integer for which this holds for $p > 1$. A matrix A is said to be *primitive* if and only if $A^m = J$ for some m.

It is clear, therefore, that for a convergent matrix A all matrix elements $a_{ij}^{(k)}$ must have limiting values. In the powers of periodic matrix, however, the matrix elements of some subset of all matrix elements must oscillate in value and the remaining elements (if any) may possibly have limiting values. Therefore, we will for convenience refer to any matrix A^l which appears infinitely often in the sequence of powers of a matrix as a *limit matrix* [321].

The number of distinct matrices in the sequence of powers of a matrix $A \times B_n$ is at most 2^r where $r = n^2$. Thus A must be either finitely convergent or finitely oscillatory. The possibilities are obviously exclusive [321].

LEMMA 5.2.1 [321]. A convergent Boolean matrix converges to an idempotent matrix.

Proof. Immediate.

REMARK 5.2.1 [321]. Convergent matrices fall into three classes: (i) primitive matrices, which converge to universal matrix, (ii) nilpotent matrices, which converge to zero matrix, (iii) matrices which converge to an idempotent other than zero matrix or universal matrix.

We need the following result about arbitrary finite cyclic semigroups proved in A. H. Clifford and G. B. Preston [56].

THEOREM 5.2.2 [56]. Let S be a finite cyclic semigroup with generator a. Let K be the least positive integer such that $a^k = a^{k+m}$ for some positive integer m. Let d be the least positive integer such that $a^k = a^{k+d}$. Then $\{a^x : x \geq k\}$ is a cyclic group of order d. If a^e is the identity element of this group, then d divides e.

Next we prove a result which was proved essentially by David Rosenblatt [321] in graph-theoretic terms, by S. Schwarz [345] in relation-theoretic terms and probably by others.

DEFINITION 5.2.2. The *period of oscillation* of a Boolean matrix A is the least positive integer d such that $A^k = A^{k+d}$ for some integer k. The least integer k such that $A^{k+d} = A^k$ for some positive integer d is called the *index* of A.

DEFINITION 5.2.3. The matrix $A \in B_n$ is said to be *decomposable* if there exists $P \in P_n$ such that

$$PAP^T = \begin{bmatrix} B & 0 \\ C & D \end{bmatrix}$$

where B and D are square; otherwise A is *indecomposable*. Now A is said to be *partly decomposable* if there exist $P, Q \in P_n$ such that

$$PAQ = \begin{bmatrix} B & 0 \\ C & D \end{bmatrix}$$

where B and D are square; otherwise A is *fully indecomposable* (see also Definition 2.4.4).

REMARK 5.2.2. Note that a matrix $A \in B_n$ is indecomposable if and only if there is no proper nonempty subset $\underline{m}$ of $\underline{n}$ such that $a_{ij} = 0$ whenever $i \in \underline{m}$, $j \in \underline{n} \setminus \underline{m}$. If A is periodic then A is either decomposable or partly decomposable. (This follows from Theorem 5.2.7).

PROPOSITION 5.2.3. A Boolean matrix A is indecomposable if and only if for any $i, j \in \underline{n}$ there exist some sequence $i_1, \ldots, i_k$ such that $a_{ii_1} = 1$, $a_{i_1 i_2} = 1, \ldots, a_{i_k j} = 1$. (Including the empty sequence with $a_{ij} = 1$).

Proof. Suppose this condition holds and suppose that $\underline{m}$ is such that $a_{ij} = 0$ for $i \in \underline{m}$, $j \in \underline{n} \setminus \underline{m}$. Assuming $\underline{m}$, $\underline{n} \setminus \underline{m}$ are nonempty, let $x \in \underline{m}$, $y \in \underline{n} \setminus \underline{m}$. Let $i_1, \ldots, i_k$ be a chain such that a_{xi_1}, $\ldots, a_{i_k y}$ all equal 1. Let i_j be the last member of the chain in $\underline{m}$. Then $a_{i_j i_{j+1}} = 1$ but $i_j \in \underline{m}$, $i_{j+1} \notin \underline{m}$. This is a contradiction.

Suppose there is no such subset $\underline{m}$. Fix x. Let S be the set of all $y \in \underline{n}$ such that there exists a chain $i_1, \ldots, i_k$ with $a_{xi_1}, \ldots,$ $a_{i_k y} = 1$.

If $S \neq \underline{n}$, let $z \notin S$. If $S = \emptyset$, then $\underline{m} = \{x\}$ gives a decomposition of A. So assume $S \neq \emptyset$. Suppose there exists $w \in S$ such that $a_{wz} = 1$. Then let $i_1, \ldots, i_k$ be a sequence from x to w with $a_{xi_1} = \ldots a_{i_k w}$ $= 1$. Thus $z \in S$, which is a contradiction. Thus for any $w \in S$, $z \notin$ S, $a_{wz} = \emptyset$. So S gives a decomposition of A. This contradicts our assumption that no $\underline{m}$ exists. Thus all x, $S = \underline{n}$. Thus the condition of the proposition is true. □

DEFINITION 5.2.4. If $A \in B_n$, and $x, y \in \underline{n}$, a sequence $i_1, \ldots,$ i_r where $i_j \in \underline{n}$ is a *connecting sequence* from x to y for A if $a_{xi_1} =$ $a_{i_j i_{j+1}} = a_{i_r y} = 1$ for $j = 1$ to r. Then the length of the connecting sequence is said to be r + 1. If $a_{xy} = 1$ we consider $\emptyset$ to be a connecting sequence of length 1 from x to y.

LEMMA 5.2.4. Let A^e be an idempotent power of an indecomposable Boolean matrix A. Then $A^e \geq I$.

Proof. Let $i \neq j$. Let s_1, s_2 be connecting sequences from i to j, j to i. Then s_1 followed by i followed by s_2 gives a connecting sequence from j to j. Let its length be m. Then the existence of this connecting sequence implies that $a_{jj}^{(m)} = 1$. Thus $a_{jj}^{(em)} = 1$. But $A^{em} = A^e$ if A^e is idempotent. Thus $a_{jj}^{(e)} = 1$ for any j. □

LEMMA 5.2.5. Let k be the index of an indecomposable Boolean matrix A. Let t be an integer greater than or equal to k. If for some t, $a_{ii}^{(t)} = 1$ then $a_{jj}^{(t)} = 1$ for every j.

Proof. Suppose $a_{ii}^{(t)} = 1$, $t \geq k$. We find a connecting sequence s, from 1 to 1 such that every number in n occurs somewhare in s. To do this first find a connecting sequence from 1 to 2, then one from 2 to 3, and so on, finally one from n to 1. Then just join these sequence together to form s. Let r be the length of s.

Let A^e be an idempotent power of A. Consider the following connecting sequence of length (er + t). Join together e copies of s and at a place where i occurs in s insert a length t connecting sequence from i to i. Such a sequence will exist if $a_{ii}^{(t)} = 1$.

We can obtain a length (er + t) connecting sequence from j to j for any j as follows. Since j occurs in s we may write s as s_1, j, s_2. Then take the rearranged connecting sequence s_2, i, s_1. This connects j to j.

Thus $a_{jj}^{(er+t)} = 1$. But since A^e is idempotent and t is greater than or equal to the index of A, $A^{(er+t)} = A^t$. □

LEMMA 5.2.6. Let A be a Boolean matrix. Let k and t be as in the statement of Lemma 5.2.5. If $A^t \geq I$ then A^t is idempotent.

Proof. By Theorem 5.2.2, for $t \geq k$, A^t lies in a cyclic subgroup. Thus there is some m such that $A^{tm} = A^t$. But if $A^t \geq I$, $A^t \leq A^{2t} \leq \dots \leq A^{mt}$. Thus all these inequalities must be equalities. Thus $A^t = A^{2t}$. □

NOTATION 5.2.1. The abbreviation g.c.d. (l.c.m.) will be used for greatest common divisor (least common multiple).

LEMMA 5.2.7. The g.c.d. of $\{x: x > 0, a_{ii}^{(x)} = 1 \text{ for some } i\}$ equals the g.c.d. of $\{x: x \geq k, a_{ii}^{(x)} = 1 \text{ for some } i\}$, if A is an indecomposable Boolean matrix and k is the index of A.

Proof. Let A^e be an idempotent power of A. If $a_{ii}^{(x)} = 1$, then $a_{ii}^{(x+e)} = 1$, with $x + e > e \geq k$. And g.c.d. $\{x, \dots, e, \dots\}$ = g.c.d. $\{x + e, \dots, e, \dots\}$. □

LEMMA 5.2.8. Let A be indecomposable Boolean matrix. Let d be the period of oscillation of A. For $y = 0, 1, \dots, d - 1$, let S_y be the set of numbers x such that there exists a connecting sequence from 1 to x of length congruent to y modulo d. Then the S_y form a partition of $\underline{n}$, and if $a_{ij} = 1$ and $i \in S_y$ then $j \in S_z$ where $z \equiv y + 1 \pmod{d}$.

Proof. Indecomposability of A implies every member of $\underline{n}$ is in an S_y. Suppose $S_y \cap S_w \neq \emptyset$ for $y \neq w$. Let $i \in S_y \cap S_w$. Let s_1 be a connecting sequence from 1 to i of length $t_1 \equiv y \pmod{d}$ and let s_2 be a connecting sequence of length $t_2 \equiv w \pmod{d}$, from 1 to i. Let s_3 be a connecting sequence of length $t_3 \geq k$, from i to 1, where k is the index of A. Then s_1, i, s_3 and s_2, i, s_3 are connecting sequence of length $t_1 + t_3$, $t_2 + t_3$ from 1 to 1. Thus $a_{11}^{(t_1+t_3)} = a_{11}^{(t_2+t_3)} = 1$. Thus by Lemma 5.2.5 and Lemma 5.2.6, $A^{(t_1+t_3)}$ and $A^{(t_2+t_3)}$ are idempotent. Thus by Theorem 5.2.2, $d | t_1 + t_3$ and $d | t_2 + t_3$. But t_2 and t_1 are not congruent modulo d, this is a contradiction. Thus $S_y \cap S_w = \emptyset$ for $y \neq w$, and the S_y partition $\underline{n}$.

If $a_{ij} = 1$ and $i \in S_y$ we can find a connecting sequence s from 1 to i of length some number $m \equiv y \pmod d$. Then adding i to this gives a connecting sequence from 1 to j of length m + 1. □

THEOREM 5.2.9 [321, 345]. Let A be an indecomposable Boolean matrix. The period of oscillation d of A is the g.c.d. of all integers x such that $a_{ii}^{(x)} = 1$ for some i. There exists a permutation matrix P such that PAP^T can be written in the form

$$\begin{bmatrix} 0 & A_1 & 0 \ldots 0 \\ 0 & 0 & A_2 \ldots 0 \\ 0 & 0 & 0 \ldots 0 \\ & \ldots\ldots\ldots\ldots & \\ A_d & 0 & 0 \ldots 0 \end{bmatrix}$$

Here the A_i's are certain matrices, the zeros are identically zero matrices, and the partitioning on the rows is identical with the partitioning on the columns.

Proof. The first statement is implied by Lemma 5.2.7, together with Lemmas 5.2.4, 5.2.5, 5.2.6.

Since S_y is a partition of $\underline{n}$ we can choose a permutation π such that π applied to 1, 2, ... , $|S_0|$ gives the set S_0, π applied to $|S_0| + 1, \ldots, |S_0| + |S_1|$ gives the set S_1, and so on, where S_y is as in Lemma 5.2.8. Then let PAP^T be partitioned by the sets 1, 2, ... , $|S_0|$, $|S_0| + 1, \ldots, |S_0| + |S_1|$, etc. We obtain the stated form. In fact $(PAP^T)_{ij} = 1$ only if $i\pi \in S_y$, $j\pi \in S_w$, $w \equiv y + 1 \pmod d$ for some y, w. This is true if and only if i, j is in one of the locations where an A_h $(1 \le h \le d)$ occurs in the form above. □

Note that the period of oscillation of A can be determined from the corresponding graph G_A. The g.c.d. mentioned in the preceding theorem is the g.c.d. of the lengths of all closed walks in G_A, i.e., sequences $v_0, v_1, \ldots, v_r$ such that $i = 0, 1, \ldots, r - 1$ there is an edge from v_i to v_{i+1}, with $v_0 = v_r$. And instead of looking at every closed walk we need only consider those which are minimal with all vertices v_0, v_r distinct. The reason is that every closed walk can be decomposed into minimal ones by removing closed subwalks.

If A is decomposable, one can obtain its period of oscillation as the l.c.m. of the periods of oscillation of certain indecomposable submatrices. These will be the matrices of the strong components of G_A.

DEFINITION 5.2.5. A *cycle* in a graph is a sequence of vertices $v_0, v_1, \ldots, v_r$ such that for $i = 0, 1, \ldots, r - 1$, there is a directed edge from v_i to v_{i+1} and $v_0 = v_r$, and all the other v's are distinct.

THEOREM 5.2.1 [321]. If $A \in B_n$ with its corresponding graph G_A, then A converges to 0 if and only if G_A contains no cycles.

Proof. In general the (i, j)-entry of A^r is 1 if and only if there is an edge sequence of length exactly r from vertex i to vertex j in G_A. If there are no cycles in G_A the vertices in each edge sequence must be distinct. Thus no edge sequence can have length larger than $n - 1$. Thus $A^n = 0$.

If there is some cycle, repetition of this cycle gives edge sequence of arbitrary large length, and no power of A will be 0. □

In the following we shall present a matrix-theoretic characterization of the asymptotic forms of powers of matrices. Most of the following results are contained in K. H. Kim [173].

DEFINITION 5.2.6. A matrix $PAP^T \in B_n$ is in *triangular form* (*normal form*), where $P \in P_n$, if:

$$\begin{bmatrix} A_{11} & 0 & \cdots & 0 \\ A_{21} & A_{22} & \cdots & 0 \\ & \cdots\cdots & & \\ A_{k1} & A_{k2} & \cdots & A_{kk} \end{bmatrix}$$

where each main diagonal block is square and $k \geq 2$. The submatrices A_{ii} are called the *constituents* of A [96].

G. Keri [172] has developed an algorithm for reducing matrices to block triangular form by multiplying them by permutation matrices. For instance this could be useful for matrices with few nonzero entries.

Let M be a matrix over $\mathcal{R}$. Let A be the Boolean matrix $a_{ij} = 0$ or 1 according as $m_{ij} = 0$ or $m_{ij} \neq 0$. Let B be the Boolean matrix of Boolean minors of A. Determine the row counts of B. Find a row with least row count among the rows of B. Find all rows equal to it. Delete the rows and columns of 1 entries in this set or rows, and repeat the process. The locations of the deleted elements give permutations P, Q for reducing B to block triangular form.

One difficulty is to find B. However as G. Keri observes, B is close to the Boolean matrix for M^{-1}, which may be used instead to aid in this process.

THEOREM 5.2.11 [173]. If $A \in B_n$ is an k-block triangular matrix and at least one constituent of A is equal to a non-identity permutation matrix, then A is periodic.

Proof. The rth power of A will have among its constituents the rth power of the permutation matrix in question. Since this part of A^r will never become constant, A^r as a whole will not converge. □

DEFINITION 5.2.7. A matrix $A \in B_n$ will be called *k-block partitionable* if there exists $P \in P_n$ such that:

$$\begin{bmatrix} A_{11} & A_{12} & \cdots & A_{1k} \\ A_{21} & A_{22} & \cdots & A_{2k} \\ & \cdots\cdots & & \\ A_{k1} & A_{k2} & \cdots & A_{kk} \end{bmatrix}$$

where each constituent is square $k > 0$.

DEFINITION 5.2.8. A matrix $A \in B_n$ will be called *row nonempty* (*column nonempty*) if every row (column) of A contains at least one 1.

THEOREM 5.2.12 [173]. Let $A \in B_n$ be a k-block partitionable matrix. If there exists a periodic matrx B of order $m > 2$ such that $b_{ij} = 1$ if $A_{ij} \neq 0$, $b_{ij} = 0$ if $A_{ij} = 0$ and either all nonzero submatrices A_{ij} are row nonempty matrices or all nonzero submatrices A_{ij} are column nonempty matrices, then A is periodic.

Proof. Assume all nonzero submatrices A_{ij} are row nonempty matrices. We will establish the following inductive hypothesis: A has the property that

$$b_{ij}^{(t)} = \begin{cases} 1 \text{ if } (A^t)_{ij} \text{ is row nonempty} \\ 0 \text{ if } (A^t)_{ij} = 0 \end{cases}$$

where t is a positive integer.

This hypothesis is true when $t = 1$. Assume it is true for $t = m$. Then $b_{ij}^{(m+1)} = \Sigma\, b_{ik}^{(m)} b_{kj}$, $(A^{m+1})_{ij} = \Sigma\, (A^m)_{ik} A_{kj}$.

If $b_{ij}^{(m+1)} = 0$, then for every k either $b_{ik}^{(m)} = 0$ or $b_{kj} = 0$, so for every k either $(A^m)_{ik} = 0$ or $A_{kj} = 0$ by the inductive hypothesis. Thus $(A^{m+1})_{ij} = 0$.

Let $M = ST$ be a product of row nonempty Boolean matrices. For any i, the ith row of S will contain a 1 in some position (i, j). Then $m_{ik} \geq s_{ij} t_{jk} > 0$. Thus for any i, $M_{i*} \neq 0$, and so M is a row nonempty matrix.

Suppose $b_{ij}^{(m+1)} = 1$ so that for some k both $b_{ik}^{(m)} = 1$ and $b_{kj} = 1$.

Then $(A^m)_{ik}$ and A_{kj} are row nonempty matrices so $(A^m)_{ij} \geq (A^m)_{ik}A_{kj}$ will be a row nonempty matrix. This completes the induction. Since B is not eventually constant, neither A will be constant. □

PROPOSITION 5.2.13 [173]. If $A \in B_n$ is decomposable and non-oscillatory, then A converges to some idempotent matrix E such that $0 \leq E < J$.

Proof. If A is non-oscillatory, then it has a limit L. If $A^t = L$, $A^{t+1} = A^{t+2} = \ldots = A^{2t} = L^2$ where t is a positive integer, and so L is idempotent matrix. If A is decomposable, then the same decomposition will apply to L. Thus $L < J$. □

PROPOSITION 5.2.14 [173]. If $A \in B_n$, then A converges to 0 if and only if there exists $P \in P_n$ such that PAP^T is equal to a subtriangular or strictly lower triangular matrix.

Proof. Exercise.

DEFINITION 5.2.9. A matrix $A \in B_n$ will be called a *row nonempty and block permutation matrix* if (i) A is k-block partitionable, and (ii) there exists a non-identity permutation matrix P of order $m \geq 2$ such that $p_{ij} = 1$ if A_{ij} is row nonempty matrix, and $p_{ij} = 0$ if $A_{ij} = 0$.

PROPOSITION 5.2.15 [173]. If $A \in B_n$ is a row nonempty matrix, then A converges to J if and only if A is neither decomposable nor a row nonempty and block permutation matrix.

Proof. Necessity. If A is decomposable, then all powers of A will be decomposable and so A cannot converge to J. If A is a row nonempty and block permutation matrix, then A is periodic by Theorem 5.2.12.

Sufficiency. Assume A is indecomposable. If the period of oscillation were greater than 1, then A is k-block partitionable such that

$$\begin{bmatrix} 0 & A_{12} & 0 & \cdots & 0 \\ 0 & 0 & A_{23} & \cdots & 0 \\ & \cdots & \cdots & \cdots & \\ 0 & 0 & 0 & \cdots & A_{k-1,k} \\ A_{k1} & 0 & 0 & \cdots & 0 \end{bmatrix}$$

by Theorem 5.2.9. Since A is indecomposable this matrix would have to be a row nonempty matrix. Thus a Boolean matrix which is indecomposable and not a row nonempty and block permutation matrix have period of oscillation 1, and so converges to J. □

Let us now rephrase the above results in the language of nonnegative matrices.

NOTATION 5.2.2. To any nonnegative square matrix A of order n we may associate in an obvious way a Boolean matrix B of order n by writing $b_{ij} = 1$ if $a_{ij} > 0$ otherwise $b_{ij} = 0$. In this representation, it is clear that for all finite powers A^p of A and B^p of B: $b_{ij}^{(p)} = 1$ if and only if $a_{ij}^{(p)} > 0$.

DEFINITION 5.2.10. If A is a nonnegative square matrix of order n with Boolean matrix representation B, then $G_B(A)$ is said to be the *graph* of A.

DEFINITION 5.2.11 [321]. A subgraph H of graph G is a *cyclic net* of order m in G if and only if H contains m > 0 points of G and each point of H is connected to every point of H.

THEOREM 5.2.16 [321]. If A is a nonnegative square matrix of order n ≥ 2, then A is indecomposable if and only if the graph $G_B(A)$ is a cyclic net.

Proof. Directly, for if we associate the index set of the points $G_B(A)$ with row and column indices of A, it is clear that A is indecomposable if and only if each point of $G_B(A)$ is connected to every point of $G_B(A)$, and so $G_B(A)$ is consequently a cyclic net. □

In passing we remark that the correspondence of concepts is readily evident in the canonical form for nonnegative indecomposable square matrices.

DEFINITION 5.2.12. Two Boolean (nonnegative) matrices A and B of same order are said to have the same *zero pattern* if $a_{ij} = 0$ whenever $b_{ij} = 0$, and vice versa.

DEFINITIOIN 5.2.13. A nonnegative square matrix that has the same zero pattern as a row nonempty matrix (row nonempty and block permutation matrix) will be called a *generalized row nonempty matrix* (*generalized row nonempty and block permutation matrix*).

THEOREM 5.2.17 [173]. If A is nonnegative square matrix of order $n \geq 2$ and is a generalized row nonempty matrix, then A is primitive if and only if A is neither decomposable nor a generalized row nonempty and block permutation matrix.

Proof. The proof follows from the fact that there is an homomorphism between Boolean matrices and nonnegative matrices of the same order. □

THEOREM 5.2.18 [319]. Let A be any Boolean matrix. Then there exists a permutation matrix P and a partitioning of PAP^T (the same on rows as on columns) such that the diagonal blocks of A are either indecomposable or 0, and the blocks above the main diagonal are 0. Suppose A is periodic, and the (i, i) or (j, j)-block of A is nonzero. Then (i, j)-block of $A^m = (A^{m-1})_{ii}A_{ij}$ or $A^m = A_{ij}(A^{m-1})_{jj}$ respectively. If A is periodic and indecomposable then there exists a permutation matrix Q and a partitioning of QAQ^T (the same on rows as on columns) such that QAQ^T has the form

$$\begin{bmatrix} 0 & J_1 & 0 & \cdots & 0 \\ 0 & 0 & J_2 & \cdots & 0 \\ 0 & 0 & 0 & \cdots & 0 \\ & & \cdots\cdots & & \\ J_d & 0 & 0 & \cdots & 0 \end{bmatrix}$$

where the J_d are universal matrices.

Proof. The matrix $(I + A)^n$ is idempotent and reflexive and so is the matrix of a quasi-order relation. The matrix of a quasi-order relation can be put into the required form by refining the quasi-order to a linear order on the same equivalence classes. And if P is a permutation matrix putting $(I + A)^n$ into the required form, P will also put into the required form. This proves the first statement.

Let A be periodic, and let the (i, i)-block of A be nonzero. For any m, let k be such that $A^{m+k} = A$. Then we have $(A^m)_{ij} \geq (A^{m-1})_{ii}A_{ij}$, $A_{ij} \geq (A^k)_{ii}(A^m)_{ij}$. By these inequalities $(A^m)_{ij} \geq (A^{k+m-1})_{ii}(A^m)_{ij}$ $\geq (A^{x(k+m-1)})_{ii}(A^m)_{ij}$ for any $x > 1$, where the second inequality follows from the first by induction. Since A_{ii} is nonzero, by the first statement of the theorem it will be indecomposable. Thus $(A^x)_{ii} \geq I$ for some $x > 1$. This implies the inequalities above are equalities. This proves the second statement.

Let A be periodic of period d and indecomposable. By Theorem 5.2.9, A can be written in the form

$$\begin{bmatrix} 0 & A_1 & 0 & \cdots & 0 \\ 0 & 0 & A_2 & \cdots & 0 \\ 0 & 0 & 0 & \cdots & 0 \\ & & \cdots\cdots & & \\ A_d & 0 & 0 & \cdots & 0 \end{bmatrix}$$

where the ith block represents the vertices of G_A reachable by an edge sequence of length congruent to i(mod d) from a vertex v_0 in the dth block, where G_A is a corresponding graph of A. Since A has period d, $A^d = A^d + A^{2d} + \ldots$. The right-hand side will not have zero entries in its main diagonal blocks, since any two vertices in the same block can be joined by an edge of length divisible by d (they can be joined some edge sequence by indecomposability and if its length were not divisible by d they would lie in different blocks). Thus A^d is a direct sum of universal matrices of different dimensions. This implies that $A = A^d A A^d$ will have the required form. □

In the following we present the primitivity of special Boolean matrices due to K. H. Kim and J. R. Krabill [181, 182] and S. Schwarz [342].

As we have seen in the Definition 1.6.3, every circulant C can be written in the form $C = c_0 I + c_1 P + c_2 P^2 + \ldots + c_{n-1} P^{n-1}$. Omitting those $c_i \in \beta_0$ and defining $P^0 + I$, we have

$$C = P^{i_1} + P^{i_2} + \ldots + P^{i_m}$$

where $0 \leq i_1 < \ldots < i_m \leq n - 1$.

THEOREM 5.2.19 [182, 342]. A circulant $C \in B_n$ is primitive if and only if g.c.d. $(i_2 - i_1, i_3 - i_1, \ldots, i_m - i_1, n) = 1$.

Proof. Write $C = P^{i_1}(I + P^{i_2 - i_1} + \ldots + P^{i_m - i_1}) = P^{i_1} T$, where T has an obvious meaning. We have

$$C^j = P^{j i_1} T^j$$

Since the permutation matrix $P^{j i_1}$ rearranges only the rows and columns in T^j, we conclude that $C^j = J$ holds if and only if $T^j = J$ holds.

Since $I \leq T$, T is primitive if and only if $J = T + T^2 + \ldots + T^n$. It is advantageous to write the above inequality instead as

$$\sum_{k=1}^{r} T^k = J$$

for any integer $r \geq n$. Hence T is primitive if and only if for any $r \geq n$ we have

$$\sum_{k=1}^{r} (I + P^{i_2 - i_1} + \dots + P^{i_m - i_1})^k = J \tag{5.2.1}$$

Note that $I + P + \dots + P^{n-1} = J$ and each summand on the left-hand side is essential, i.e., omitting any P^i $(0 \leq i \leq n - 1)$ the sum becomes not equal to J.

Multiply term by term the product $(I + P^{i_2 - i_1} + \dots + P^{i_m - i_1})^k$. Using the idempotency of addition and $P^n = I$, the left-hand side of (5.2.1) finally becomes a sum of distinct powers of P. Now (5.2.1) holds if and only if the left-hand side of (5.2.1) contains as a summand every power P^k for $k = 0, 1, \dots, n - 1$. Since this expression certainly contains I, we can state that (5.2.1) holds if and only if to any integers $t = 1, 2, \dots, n - 1$ there exist nonnegative integers $x_{2k}, x_{3k}, \dots, x_{mk}$ such that $x_{2k}(i_2 - i_1) + x_{3k}(i_3 - i_1) + \dots + x_{mk}(i_m - i_1) \equiv t \pmod n$. Now the congruence $x_2(i_2 - i_1) + x_3(i_3 - i_1) + \dots + x_m(i_m - i_1) \equiv 1 \pmod n$ has a solution $x_{21}, x_{31}, \dots, x_{m1}$ if and only if g.c.d. $(i_2 - i_1, i_3 - i_1, \dots, i_m - i_1, n) = 1$. On the other side if this condition is satisfied, then for any $t = 2, 3, \dots, n - 1$ the congruence $y_2(i_2 - i_1) + y_3(i_3 - i_1) + \dots + y_m(i_m - i_1) \equiv t \pmod n$ has a solution $y_{2t}, y_{3t}, \dots, y_{mt}$. (It is sufficient to put $y_{2t} = tx_{21}, \dots, y_{mt} = tx_{m1}$). This proves the theorem. □

C.-Y. Chao and S. Winograd [50] generalized the above result to companion matrices.

THEOREM 5.2.20 [50]. Let Boolean matrix

$$\begin{bmatrix} 0 & 0 & 0 \dots 0 & b_0 \\ 1 & 0 & 0 \dots 0 & b_1 \\ 0 & 1 & 0 \dots 0 & b_2 \\ & \dots\dots\dots & & \\ 0 & 0 & 0 \dots 1 & b_{n-1} \end{bmatrix}$$

Let $0 \le j_1 < \dots < j_t \le n - 1$ be the set of locations of 1 entries in the last column of C. Let $A = C^{i_1} + C^{i_2} + \dots + C^{i_k}$. Then A is primitive if and only if $b_0 = 1$ and g.c.d. $(i_2 - i_1, \dots, i_k - i_1, j_2 - 1, \dots, j_t - 1) = 1$.

Proof. Straightforward. □

THEOREM 5.2.21 [49]. Let $n \ge 5$ be odd and let $A \in B_n$ be the matrix such that $a_{12} = a_{23} = \dots = a_{m-1,m} = a_{1m} = 1$ and $a_{2,m+1} = a_{m+1,m+2} = \dots = a_{2m-4,2m-3} = a_{2m-3,1} = 1$ and all other entries are 0, where $n = 2m - 3$. Then none of $A, A^2, \dots, A^n$ is fully indecomposable.

Proof. Let v_1 denote the vector which has a 1 in place 4 and only there. Let v_2 denote the vector which has a 1 entry in place $m + 1$ and only there.

Then v_1A^i for $i = 1$ to n is the series of vectors whose 1 entries are in locations 5, ... , 1, 2, {3, m + 1}, ... , {1, 2}. And v_2A^i for $i = 1$ to n is the series of vectors whose 1 entries are in locations m + 2, ... , 1, 2, {3, m + 1}, ... , {1, 2}. Thus for $i = 1$ to n the weight of $(v_1 + v_2)A^i$ is less than or equal to the weight of $(v_1 + v_2)$. Thus $A, A^2, \dots, A^n$ are not fully indecomposable. □

C.-Y. Chao [49] gives a similar construction for n even and greater than 5.

5.3 Primitive Matrices

In this section we study primitive matrices in more detail. We first show that as $n \to \infty$ that probability that a random Boolean matrix is fully indecomposable tends to 1. Since every fully indecomposable matrix is primitive, this establishes that most large Boolean matrices are primitive. Counting the number of Hall matrices is an unsolved problem, but our method yields an estimate.

Next we prove a series of results leading to Theorem 5.3.11 which states that every nonsingular Boolean matrix of nonzero permanent, and also every idempotent Boolean matrix with a 1 in each row and column, is a product of primitive matrices. In the process we derive another important result about idempotent matrices (Lemma 5.3.9).

Minimal fully indecomposable matrices are called nearly decomposable matrices. In Theorem 5.3.12 we summarize much of what is known about these matrices.

DEFINITION 5.3.1. A matrix $A \in B_n$ is said to be *r-decomposable* if and only if there are nonempty sets $\underline{s}, \underline{t} \subset \underline{n}$ such that $a_{ij} = 0$ whenever $i \in \underline{s}$, $j \in \underline{t}$ and $|\underline{s}| + |\underline{t}| = n + 1 - r$. Clearly, a *0-decomposable* matrix is a fully indecomposable matrix. Let Dec (B_n^r) denote the set of all r-decomposable matrices in B_n.

A result similar to the theorem below was obtained by P. Erdös and A. Renyi [103] in 1968.

THEOREM 5.3.1 [206].

$$2n\sum_{i=0}^{r} \binom{n}{i} 2^{n^2-n} + 2\binom{n}{2}\binom{n}{r+1} 2^{n^2-2(n-r-1)} + 2\binom{2n}{n-r+1} 2^{n^2-3(n-r-2)} \geq |\text{Dec } (B_n^r)| \geq 2\binom{n}{1}\sum_{i=0}^{r} \binom{n}{i} 2^{n^2-n} -$$

$$2\binom{n}{2}\sum_{j=0}^{r}\sum_{i=0}^{r} \binom{n}{i}\binom{n}{j} 2^{n^2-2n} - n^2 \sum_{i=0}^{r}\sum_{j=0}^{r} \binom{n}{i}\binom{n}{j} 2^{n^2-2n+1}$$

Proof. Let M(S, T) be the set of all matrices $A \in B_n$ which have $a_{ij} = 0$ whenever $i \in S \subset \underline{n}$, $j \in T \subset \underline{n}$ with $|S| + |T| = n - r + 1$. Thus $|M(S, T)| = 2^{n^2-|S||T|}$. Thus Dec (B_n^r) is the union of all sets M(S, T)

We will group the M(S, T) according to the numbers of elements in S and T.

Case 1. $|S| = 1$ or $|T| = 1$

Case 2. $|S| = 2$ or $|T| = 2$, and neither is 1

Case 3. $|S| \geq 3$ and $|T| \geq 3$

In Case 1 consider first the sets S, T with $|S| = 1$. For a fixed k, $\bigcup_{S=\{k\}} M(S, T)$ is the set of all matrices of order n whose kth row contains at least n - r zeros. If this row has i 1's, for $i \le r$, we can distribute them in $\binom{n}{i}$ ways. We can then choose the other entries in 2^{n^2-n} ways. So

$$\left| \bigcup_{S=\{k\}} M(S, T) \right| = \sum_{i=0}^{r} \binom{n}{i} 2^{n^2-n}$$

Therefore

$$\left| \bigcup_{|S|=1} M(S, T) \right| \le n \sum_{i=0}^{r} \binom{n}{i} 2^{n^2-n}$$

By symmetry

$$\left| \bigcup_{|T|=1} M(S, T) \right| \le n \sum_{i=0}^{r} \binom{n}{i} 2^{n^2-n}$$

In Case 2, the sets S, T can be chosen in $\binom{n}{2}\binom{n}{r+1}$ ways for $|T| = 2$. Thus

$$\left| \bigcup_{\text{Case } 2} M(S, T) \right| \le 2\binom{n}{2}\binom{n}{r+1} 2^{n^2-2(n-r-1)}$$

In Case 3, $|M(S, T)|$ will be less than or equal to $2^{n^2-3(n-r-2)}$. The number of ways to choose the set S, T is less than

$$\sum_{j=0}^{r} \binom{n}{j}\binom{n}{n+1-r-j} = \binom{2n}{n+1-r}$$

Therefore

$$\left| \bigcup_{\text{Case } 3} M(S, T) \right| \le \binom{2n}{n+1-r} 2^{n^2-3(n-r-2)}$$

This proves the first inequality.

To prove the second inequality we consider only Case 1. Let

$$M_k = \left| \bigcup_{S=\{k\}} M(S, T) \right|, \quad N_k = \left| \bigcup_{T=\{k\}} M(S, T) \right|$$

Then

$$|M_k| = \sum_{i=0}^{r} \binom{n}{i} 2^{n^2-n} = |N_k|$$

For $p \ne q$,

$$|M_p \cap M_q| = \sum_{i=0}^{r} \sum_{j=0}^{r} \binom{n}{i}\binom{n}{j} 2^{n^2-2n}$$

by first choosing the 1's in the pth row, then the 1's in the qth column, then all other entries. Thus by the first two terms of the inclusion-exclusion formula, the second inequality holds. (These two terms will count no element more than once). The last term of the inequality comes similarly from $|M_p \cap N_q|$. □

REMARK 5.3.1. Notice from Definition 5.3.1 that the concept of r-decomposability is the same as in the case of (0, 1)-matrices over $\mathcal{R}$. Thus, the asymptotic result in Theorem 5.3.1 holds for n × n (0, 1)-matrices over $\mathcal{R}$

COROLLARY 5.3.2. The number of r-decomposable Boolean matrices, for fixed r, is asymptotically equal to

$$2n\binom{n}{r}2^{n^2-n}$$

COROLLARY 5.3.3. The number of r-indecomposable Boolean matrices, for fixed r, is

$$2^{n^2}\left(1 - o\left(\frac{n^{r+1}}{2^n}\right)\right)$$

COROLLARY 5.3.4. The proportion of types of nonnegative matrices which are fully indecomposable approaches 1 as $n \to \infty$.

COROLLARY 5.3.5. The proportion of Boolean matrices which are primitive approaches 1 as $n \to \infty$.

DEFINITION 5.3.2. Let Q_n denote the set of all Boolean matrices containing a permutation and a single 1 entry not in this permutation. Then $A \in Q_n$ if and only if $PAQ \in Q_n$ for $P, Q \in P_n$.

NOTATION 5.3.1. Let E(i, j) denote the Boolean matrix whose (i, j)-entry is 1 and whose other entries are all 0.

PROPOSITION 5.3.6. If $P + E(i, j) \in Q_n$ where $P \in P_n$, then $P + E(i, j)$ is primitive if and only if (i) P is an n-cycle, (ii) $d_p(b, a) + 1$ is relatively prime to n where $d_p(b, a)$ is the directed distance from b to a in the graph of P, i.e., the least power m of P such that $bP^m = a$.

Proof. By Theorem 5.2.10, a necessary and sufficient condition for a matrix to be primitive is that its graph be strongly connected, and the greatest common divisor of the lengths of the minimal cycles in its graph be 1.

Suppose P has more than one cycle. Then an additional edge will be between points of a cycle, or lead from cycle to another. The graph of $P + E(i, j)$ then may be weakly connected, but cannot be strongly connected.

Suppose then that P is an n-cycle. There are two minimal cycles: P of order n and $a \to b \to bP \to \ldots \to a$ of order $d_p(b, a) + 1$. So the greatest common divisor of the lengths of the minimal cycles is 1 if and only if $(n, d_p(b, a) + 1)$ are relatively prime. □

REMARK 5.3.2. Note that if P is an n-cycle, $P + E(a, a)$ and $P + E(a, aP^2)$ will always be primitive.

PROPOSITION 5.3.7. A product $(P_1 + E(a, b))(P_2 + E(c, d))$ of elements of Q_n belongs to Q_n if and only if $c = aP_1$, $d = bP_2$.

Proof. For this product to belong to Q_n, $P_1E(c, d)$ must equal $E(a, b)P_2$. So $E(cP_1^T, d) = E(a, bP_2)$, and in this case $E(a, b)E(c, d)$ is either 0 or equal to both, so the product will belong to Q_n. □

THEOREM 5.3.8 [206]. Let S be the set of all products $A_1A_2 \ldots A_k$ for any k of primitive elements of Q_n such that $A_1A_2 \ldots A_k$ again belongs to Q_n. Then for n even, $S = Q_n$. For n odd, S is the set of all elements of Q_n whose permutation belongs to the alternating group of degree n.

Proof. We first study the elements T + E(a, b) where T is a 3-cycle. For the moment we assume $n \geq 6$.

Let P = (123 ... n), Q = $(1324 \ldots n)^{-1}$ = (n ... 4231). Then (P + E(5, 5))(Q + E(6, 4)) = (132) + E(5, 4) and the elements on the left-hand side are primitive. Conjugation by a permutation matrix preserves Q_n. So Q_n contains all matrices (abc) + E(d, e) where a, b, c, d, e are distinct.

For n = 5, this is still true if the 6 is replaced by a 1. Also (P + E(4, 4))(Q + E(5, 2)) = (132) + E(4, 2) and the elements on the left-hand side are primitive. Conjugation by a permutation matrix and transposes give all matrices (abc) + E(d, e) where one of d, e is equal to one of a, b, c but the other is not.

For n = 4, this is still true if the 5 is replaced by a 1. Also (P + E(2, 2))(Q + E(3, 3)) = (132) + E(2, 3) and the elements on the left-hand side are primitive. This gives all elements (abc) + E(d, e) where $d \neq e$, and d, e are among a, b, c and the permutation abc does not send d to e. Also (P + E(1, 3))(Q + E(2, 1)) = (132) + E(1, 1) and the elements on the left-hand side are primitive. This gives all matrices (abc) + E(d, d) where d is one of a, b, c. Thus for $n \geq 3$, all matrices (abc) + E(d, e) such that $d(abc) \neq e$ can be obtained, thus all matrices of this form are in Q_n.

Suppose we have a product (P + E(a, b))(Q + E(aP, bQ)). Then $aP \neq b$ if and only if $(aP)Q \neq bQ$ if and only if $a(PQ) \neq bQ$. So if E(a, b) ranges over all allowable pairs, the E term of the product will too. Thus from the 3-cycles we can obtain all matrices in Q_n whose permutations are in the alternating group.

If n is odd, since n cycles will lie in the alternating group, these are all the permutations which can be obtained. For n even and n = 4 we can obtain all elements of the form H + E(a, b) where H is an n - 1 cycle.

Let P' = (1234 ... n), Q' = $(1234 \ldots n-1)^{-1}$ = (n - 1 ... 4321). Then (P' + E(1, 1))(Q' + E(2, n - 1)) = (n(n - 1)) + E(1, n - 1), (P' + E(n, n))(Q' + E(1, n)) = (n(n - 1)) + E(n, n), (P' + E(1, 3))(Q' + E(2, 2)) = (n(n - 1)) + E(1, 2). Thus by conjugation and transposes we can obtain all matrices (ab) + E(d, e) such that $d(ab) \neq e$.

Then reasoning with these in the same way as for 3-cycles before we can obtain all elements of Q_n. For n = 2, the theorem can proved by finding the required products explicitly. □

We next summarize a number of results about idempotents.

LEMMA 5.3.9 [322]. (1) An idempotent with strongly connected graph is equal to the universal matrix.

(2) For an idempotent E, there exists a permutation matrix P such that PEP^T is a block triangular matrix in which each block consists entirely of 1's or entirely of 0's.

(3) If written in this form, either some row or greater than or equal to another, or the idempotent is the identity.

(4) If written in this form, if there are zeros on the main diagonal, or if any of the blocks are not 1×1 matrices, the rank of E is less than n.

(5) If two off-main diagonal 1's lie in the locations (a, b), (b, d) there is also an off-main diagonal 1 in location (a, d).

(6) If written in triangular form, and all the blocks are 1×1 and the idempotent is not in the identity but has at least one 1 in each row and each column then there exists a permutation matrix P such that PEP^T is still a triangular matrix and contains one of the following configurations of 1 entries:

(i, i)	
(i + 1, i)	(i + 1, i + 1)
(i, i)	
(i + 1, i)	
(j, i)	(j, i + 1), (j, j)
(j, j)	
(i, j)	
(i + 1, j)	(i + 1, i), (i + 1, i + 1)

Proof. (1) Follows from Theorem 5.2.10 and the g.c.d. of the lengths of its minimal cycles is 1.

(2) This assertion can be proved by taking as blocks the strongly connected components of the graph together with all vertices which belong to no such components regarded as blocks in themselves. Then there is a partial order on the set of blocks defined by $S_i < S_j$ if there is some directed edge from an element of S_j to one of S_i. Arrange the blocks so that for $i < j$, $S_i \not> S_j$, first choosing the minimal blocks, then those blocks connected only to minimal blocks and so on. This gives a block triangular form.

On graph-theoretic characterization of an idempotent is that if there is a directed path from i to j there is one of length 1. This implies that all blocks of the matrix consist entirely of 1's or entirely of 0's.

(3) If there is an off-main diagonal 1 in location (i, j) then $E_{i*} \geq E_{j*}$. If there are no off-main diagonal ones, the result is evidently true.

(4) It can be seen that the highest row with a main diagonal 0 is dependent on the rows above it.

(5) Follows directly from the matrix equation $E^2 = E$.

(6) Let (a, b) be any off-main diagonal 1 entry in such a matrix. Then since $E^n = E$ for all n there exist arbitrarily long sequences $(a, a_1), (a_1, a_2), \ldots, (a_{n-1}, b)$ where each pair is the location of a 1 entry in E. For large n, some pair (a_i, a_{i+1}) must have $a_i = a_{i+1}$. Then we have that E has 1's in location (a, a_i), (a_i, a_i) and (a_i, b). Here a_i may equal a or a_i may equal b.

Now in E consider those off-main diagonal 1's in locations (a, b) with the minimality property that for $c \leq a$, $d \geq b$ there is no 1 in location (c, d), unless (c, d) = (a, b) or lies on the main diagonal. If there are off-main diagonal 1's there must be one with this minimality property.

Case 1. The entries in locations (a, a) and (b, b) are 1. Conjugate E by a permutation sending $x \to x$ unless $b \leq x \leq a$; $b \to b$; $a \to b + 1$; $c \to c + 1$ for $b < c < a$. The triangular form will be preserved, and a configuration of the first type is obtained.

Case 2. The entry in location (a, a) is 1, the entry in location (b, b) is 0. Since there is at least one 1 in every row there must be a 1 entry in a location (b, c). Choose c to be a maximum. Suppose the (c, c)-entry is not 1. Then by the above there must be an x, b > x > c such that 1's exist in locations (b, x), (x, x) and (x, c). This contradicts the minimality of c. Thus the (c, c)-entry is 1. From the (a, b) and (b, c)-entries being 1, the (a, c)-entry is 1. The same conjugation as above gives a configuration of the third type.

Case 3. The entry in location (a, a) is 0, the entry in location (b, b) is 1. Then a configuration of the second type is obtained.

Case 4. The entries in location (a, a) and (b, b) are 0. Then (a, b) cannot be minimal since there will be x, a > x > b such that there are 1's in locations (a, x), (x, x) and (x, b). This proves assertion (6). This completes the proof. □

DEFINITION 5.3.3. A nonsingular Boolean matrix is *even* or *odd* according to whether the permutation it contains is even or odd.

DEFINITION 5.3.4. Let Pmt (B_n) denote the subsemigroup of B_n generated by all primitive matrices.

LEMMA 5.3.10. Pmt (B_n) is sent to itself by any of the following operations: (i) $A \to PAP^T$ for $P \in P_n$, (ii) $A \to A^T$, (iii) $A \to PAQ$ if A has some row greater than or equal to some other row, and some column greater than or equal to another column and PQ is even if n is odd, and (iv) a row operation $A_{i*} \to A_{i*}$, $A_{j*} \to A_{j*} + A_{k*}$, i = j, for some k. In addition, if the entries in a matrix of Pmt (B_n) are replaced by matrices, with a 1 entry replaced by a matrix with all entries 1 and a 0 entry replaced by a matrix with all 0's such that the result is a partitioned matrix, it will belong to the Pmt (B_n) of its dimension.

Proof. (i) and (ii) and the last statement follow by looking at the generators of Pmt (B_n), i.e., primitive matrices.

(iv) Follows from Theorem 5.3.8. The row operations of (iv) are multiplications by matrices of Q_n whose permutation is the identity.

For (iii), PAQ = P'A'Q' where P' permutes the rows as P does but then adds some row to a row greater than or equal to it, and Q' permutes the columns in the same way that Q does but then adds some column to a column greater than or equal to it. Thus P', Q' belong to Q_n. For n odd an additional conjugation by an actual permutation proves the case P and Q both odd. □

THEOREM 5.3.11 [206]. (1) A nonsingular matrix belongs to Pmt (B_n) if it is not a permutation and, if n is odd, it is even. (For n odd some odd nonsingular matrices also belong to Pmt (B_n)).

(2) An idempotent belongs to Pmt (B_n) if and only if it has a 1 in every row and every column, and is not the identity.

Proof. We prove (2) first. By conjugation we may assume the idempotent A is written in block triangular form by Lemma 5.3.9. We may assume all blocks are 1×1 and that one of the configurations of Lemma 5.3.9 (6) is in the matrix.

Let $P = (n(n-1) \ldots 1)$ be a permutation so that PA = A with its lowest row moved to the top and the other rows shifted down by 1. We will show that PA is primitive.

First we will show that the graph of PA is strongly connected. First we consider all vertices reachable from v_1 in the graph of PA. The matrix A is an idempotent in triangular form which has at least one 1 entry in each row and each column. Since there is a 1 in each column $a_{nn} = 1$. Thus in the graph of PA, v_n is reachable from v_1. Suppose $v_{n-1}, \ldots, v_k$ are reachable from v_1, where $k > 1$. In the column A_{*k-1} there will be a 1 entry. Suppose $a_{m,k-1} = 1$ where $n \geq m \geq k - 1$. Then v_{m+1} is reachable from v_1 in the graph of PA. Thus v_{k-1} will also be. So all vertices are reachable from v_1.

Next we consider all vertices from which v_1 is reachable, in the graph of PA. Since A has a 1 in each row $a_{11} = 1$. So v_1 is reachable from v_2 in the graph of PA. Suppose v_1 is reachable from any of v_k, v_{k-1}, ... , v_2. The row A_{k*} will contain a 1 entry a_{km} for some $k \geq m$. Thus in the graph of PA there is a directed path from v_{k+1} to v_m and so a directed path from v_{k+1} to v_1. Thus v_1 is reachable from any vertex other than v_1. Thus the graph of PA is strongly connected.

Finally we observe that any of the three configurations implies the g.c.d. of the lengths of all cycles is 1.

For instance with the second configuration there are two routes from v_{j+1} to v_i, one of length 1 and another of length 2. Thus PA is primitive.

Lemma 5.3.10 (iii), A is in Pmt (B_n). Now again by Lemma 5.3.10 (iii) imply that nonsingular matrices satisfying the hypothesis of assertion (i) belong to Pmt (B_n).

An example of a nonsingular matrix for n odd which is odd and primitive is

$$\begin{bmatrix} 1&0&0&0&0\\ 1&1&0&0&0\\ 1&1&1&0&0\\ 1&1&1&1&0\\ 1&1&1&1&1 \end{bmatrix} \begin{bmatrix} 0&0&0&0&1\\ 0&0&0&1&0\\ 1&0&0&0&0\\ 0&1&0&0&0\\ 0&0&1&0&0 \end{bmatrix} = \begin{bmatrix} 0&0&0&0&1\\ 0&0&0&1&1\\ 1&0&0&1&1\\ 1&1&0&1&1\\ 1&1&1&1&1 \end{bmatrix}$$

This proves the theorem. □

EXAMPLE 5.3.1. Let

$$A = \begin{bmatrix} 1&0&0&0\\ 0&1&1&0\\ 0&0&1&1\\ 0&1&0&1 \end{bmatrix}$$

Then $A \notin$ Pmt (B_4), since for any factorization of it, one of the factors will be a permutation matrix.

EXAMPLE 5.3.2. Let

$$A = \begin{bmatrix} 0&1&1\\ 1&0&0\\ 1&0&0 \end{bmatrix}$$

Then $A \not\in \text{Pmt}\ (B_3)$ since for any factorization of it by matrices with at least one 1 in every row and every column, one of the factors will be cogredient to it, and no matrix cogredient to it is primitive.

A considerable literature has developed around the topic of the nearly decomposable matrix: D. J. Hartfiel [148, 149], H. Minc [253, 254], R. Sinkhorn and P. Knopp 370 , and others.

König's theorem states that every Hall matrix is greater than or equal to some permutation matrix, i.e., that the minimal Hall matrices are permutation matrices. Minimal fully indecomposable matrices are called *nearly decomposable* matrices, and their forms are more complicated.

DEFINITION 5.3.5. A Boolean matrix A is *nearly decomposable* if and only if it is fully indecomposable and every matrix less than A is partly decomposable.

THEOREM 5.3.12 [148, 149, 253, 370]. A fully decomposable Boolean matrix is nearly decomposable if and only if for each i, either the ith row sum or the ith column sum is 2. No 2 × 2 rectangles of ones exist in any nearly decomposable matrix. The weight of a nearly decomposable n × n Boolean matrix is between 2n and 3n - 3 inclusive. Every nearly decomposable Boolean matrix is cogredient to a matrix of the form

$$
\begin{array}{c}
\\ \\ \\ \\ \\ \\ a \\ \\ \\ \\
\end{array}
\left[\begin{array}{ccccc|cccccc}
 & & & & & & & & & b & \\
1 & 1 & 0 & \cdots & 0 & 0 & 0 & 0 & \cdots & 0 & 0 \\
0 & 1 & 1 & \cdots & 0 & 0 & 0 & 0 & \cdots & 0 & 0 \\
0 & 0 & 1 & \cdots & 0 & 0 & 0 & 0 & \cdots & 0 & 0 \\
 & & \cdots & & & & & \cdots & & & \\
0 & 0 & 0 & \cdots & 1 & 0 & 0 & 0 & \cdots & 1 & 0 \\
\hline
0 & 0 & 0 & \cdots & 0 & & & & & & \\
1 & 0 & 0 & \cdots & 0 & & & & & & \\
0 & 0 & 0 & \cdots & 0 & & & H & & & \\
 & & \cdots & & & & & & & & \\
0 & 0 & 0 & \cdots & 0 & & & & & &
\end{array}\right]
$$

where H is again nearly decomposable, and unless H is 1 × 1, the (a, b)-entry of H is 0, and the row sum of H_{i*} equals 2 whenever $h_{ib} = 1$

and the column sum of H_{*j} equals 2 whenever $h_{aj} = 1$. Conversely, any such matrix is nearly decomposable.

Proof. Exercise.

DEFINITION 5.3.6. A matrix over the nonnegative real numbers is *doubly stochastic* if and only if all row and column sums are 1.

Every doubly stochastic matrix is a convex sum of permutation matrices. Thus the problem is reduced to characterizing those Boolean matrices which are sums of permutation matrices.

In the following we state a well-known theorem.

THEOREM 5.3.13. A Boolean matrix A is a sum of permutation matrices if and only if there exist permutation matrices P, Q such that PAQ is a direct sum of fully indecomposable matrices.

Proof. We first show that fully indecomposable matrices are sums of permutation matrices. This is equivalent to showing that if $a_{ij} = 1$ for a fully indecomposable matrix A then a_{ij} enters some nonzero diagonal product. This is equivalent to the fact that if we delete the ith row and jth column we get a Hall matrix. Suppose not. Then in the deleted matrix we have $r \times s$ rectangle of 0's, where $r + s = n$, by König's theorem. But then the matrix A is not fully indecomposable. This is a contradiction. Thus every fully indecomposable matrix is a sum of permutation matrices. Likewise every direct sum of fully indecomposable matrices is a sum of permutation matrices. This proves sufficiency.

Suppose A is a sum of permutation matrices. Choose permutation matrices P, Q such that PAQ is the direct sum of a maximum number of summands, and such that in each summand, we have a block form where the A_{ii} are indecomposable matrices or zero:

$$\begin{bmatrix} A_{11} & 0 & \cdots & 0 \\ * & A_{22} & \cdots & 0 \\ & \cdots\cdots\cdots & & \\ * & * & \cdots & A_{kk} \end{bmatrix}$$

That this can be done follows either by taking the strong components of the graph of A, or by using the results of S. Schwarz [345] on asymptotic forms of Boolean matrices, or just by induction, and the definition of indecomposable matrix. Since PAQ is a sum of permutation matrices, some power of PAQ will be reflexive. Thus no main diagonal block of PAQ can be zero. Let B = PAQ. If $b_{ij} = 1$ then $t_{ij} = 1$ for some permutation matrix $T \le B$. Let m be such that $T^m = T^{-1}$, where T^{-1} is an inverse of T. Then $B^m \ge T^{-1}$ so $b_{ji}^{(m)} = 1$. This implies that all blocks labelled * above are zero.

Thus PAQ is a direct sum of indecomposable matrices. Suppose some summand were not fully indecomposable. Then we could modify P, Q on this summand only, and put that summand into a form

$$\begin{bmatrix} A_{11} & 0 \\ * & A_{22} \end{bmatrix}$$

As above * will be zero. Then the original P, Q were not chosen so as to have a maximum number of summands. This is a contradiction, and completes the proof of the theorem. □

DEFINITION 5.3.7. A problem is *NP* (*nondeterministic polynomial*), roughly, if any proposed solution could be checked for validity by a polynomial-time algorithm. That a problem x is *NP-complete* means x is NP and if x could be solved by a polynomial-time algorithm, so could any NP problem. It is generally thought that NP-complete problems cannot be solved by polynomial-time algorithms.

M. Tchuente [381] has shown that the following problem is NP-complete: for a given A, B ε B_n do there exist P, Q ε P_n such that PAQ is a product of Boolean matrices less than or equal to B.

5.4 Series of Powers of Boolean Matrix

In previous section we studied primarily the period of Boolean matrices and the periodic powers. Here we focus on the index. First we study certain sums of powers related to the transitive closure. Then Proposition 5.4.7 shows that the total index and period can be computed from the index and period of the sequences e_iA^n where e_i is a vector having a 1 in place i and nowhere else. Proposition 5.4.8 shows that these new periods k_i satisfy the relationship established earlier for the overall period. Theorem 5.4.9 and Proposition 5.4.10 combine to give the major result Theorem 5.4.11 which asserts that the index of an $n \times n$ Boolean matrix is less than or equal to $(n - 1)^2 + 1$. The next result of K. H. Kim and F. W. Roush, strengthens this bound for some matrices. (See also K. H. Kim [173] where a still better result is proved). We quote a few more results of S. Schwarz without proof. In the remainder of the section we consider a generalization due to N. J. Pullman of the index in which we consider not powers of a signle matrix but products of different matrices. If the matrices commute, his Theorem 5.4.24 holds. For related work see K. H. Kim and F. W. Roush [189].

DEFINITION 5.4.1. The series of powers R, R^2, R^3, ... of a binary relation forms a semigroup and if k is the least power such that R^k is equal to infinitely many other powers of R, the set of distinct powers from R^k form a group, which is cyclic. The order of this group will be denoted d and called the *period of oscillation* of R. A generator of this group will be denoted by g, so that the elements of the groups are g, g^2, ... , g^d. The group will be denoted by Gr (R).

REMARK 5.4.1. The above definition remains true if "binary relation" is replaced by "Boolean matrix" and below we will translate

Schwarz's results into results about Boolean matrices.

DEFINITION 5.4.2. Let $A \in B_n$. The matrix $A + A^2 + A^3 + \ldots$ (sum of all distinct powers of A) is called the *transitive closure* of A. Let Tc (A) denote the transitive closure of A.

PROPOSITION 5.4.1. Let $A \in B_n$. Then $\mathrm{Tc}(A) = A + A^2 + \ldots + A^n$.

Proof. In the graph of Tc (A), two vertices i and j are connected by an edge if and only if they are connected by an edge in some A^i, i.e., by some edge sequence in the graph of A. But if so there will be some edge sequence of length less than or equal to n in the graph of A between the vertices, since we may assume the edge sequence contains each vertex at most once. So the vertices will be joined in the graph of A^i for some $i \le n$. □

DEFINITION 5.4.3. A digraph G is a W_s^t-*graph* if and only if for any two vertices v_1 and v_2 of G there is in G exactly one directed walk from v_1 to v_2 whose length c satisfies $s \le c \le t$.

This definition is equivalent to the equation $A^s + A^{s+1} + \ldots + A^t = J$ where A is the adjacency matrix of G.

PROPOSITION 5.4.2 [34]. There exists an $n \times n$ Boolean matrix such that $A^s + A^{s+1} + \ldots + A^t = J$ if and only if $n = d^s + d^{s+1} + \ldots + d^t$ for some integer d and (i) $t = s$, $d \ge 1$, $n = d^s$, or (ii) $t = s + 1$, $d \ge 1$, $n = d^s + d^t$, or (iii) $t \ge a + 2$, $d = 1$, $n = t - a + 1$, or (iv) $t \ge s = 0$, $d = 0$, $n = 1$. In (iii) and (iv) the only such graphs are G_{t-s+1}, the graph which is a single directed cycle on $t - s + 1$ vertices, and G_1, the graph with one vertex and no edges.

Proof. Exercise.

Any subset of B_n is a poset. A number of important subsets of B_n are lattices under this partial ordering, i.e., any two elements have a least upper bound and a greatest bound. The following are number of examples: reflexive matrices, symmetric matrices, transitive matrices, and any combination of these. In each case greatest lower bound is given by intersection. Least upper bound is given by the intersection of all matrices greater than or equal to the given matrix. Z. Shmuely [366] has observed that the set of all idempotent Boolean matrices is also a lattice.

THEOREM 5.4.3 [366]. For idempotents $E_1, E_2, \ldots, E_k$ in B_n, $\mathrm{Tc}\,(E_1 + \ldots + E_k)$ is idempotent and is the least upper bound of E_1, $E_2, \ldots, E_k$. The powers of the matrix $E_1 \odot \ldots \odot E_k$ converge to the greatest lower bound E of $E_1, \ldots, E_k$ where $E \in \mathrm{Idem}\,(B_n)$.

Proof. Since $\mathrm{Tc}\,(E_1 + \ldots + E_k)$ is a transitive closure, it is transitive. Thus $\mathrm{Tc}^2\,(E_1 + \ldots + E_k) \le \mathrm{Tc}\,(E_1 + \ldots + E_k)$. The formula for $\mathrm{Tc}\,(E_1 + \ldots + E_k) = \Sigma E_i + (\Sigma E_i)^2 + (\Sigma E_i)^3 + \ldots$. If we square this expression every possible expression of the form $E_{i_1} E_{i_2} \ldots E_{i_r}{}^r$ will appear. Such an expression equals $E_{i_1} E_{i_2} \ldots E_{i_r}$. Thus every term of $\mathrm{Tc}\,(E_1 + \ldots + E_k)$ is equal to a term of $\mathrm{Tc}^2\,(E_1 + \ldots + E_k)$. Thus $\mathrm{Tc}^2\,(E_1 + \ldots + E_k) \ge \mathrm{Tc}\,(E_1 + \ldots + E_k)$. It follows that $\mathrm{Tc}^2\,(E_1 + \ldots + E_k) = \mathrm{Tc}\,(E_1 + \ldots + E_k)$. From the formula, $\mathrm{Tc}\,(E_1 + \ldots + E_k) \ge E_i$ for each i. And any transitive relation which is greater than or equal to E_i must be greater than or equal to $E_1 + E_2 + \ldots + E_k$ and so also $\mathrm{Tc}\,(E_1 + \ldots + E_k)$. This proves the first statement.

An intersection of transitive matrices is transitive. Thus $F = E_1 \odot \ldots \odot E_k$ is transitive. Thus $F^2 = F$. Thus the powers of F form a descending chain $F \ge F^2 \ge F^3 \ge \ldots$. This chain must terminate, so the powers of F will all eventually equal to a matrix E. From its definition as a limit of powers, E will be idempotent.

Moreover E will be less than or equal to F and so less than or equal to E_i. Let H be any idempotent matrix less than or equal to each E_i. Then $H \le F$. Since H is idempotent, $H \le F^k$ for any k. Thus

$H \in \text{Idem}(B_n)$. This proves the theorem. □

The following results are due to S. Schwarz [345]. Some of these results were also obtained by D. Rosenblatt [321].

THEOREM 5.4.4 [345]. For any $A \in B_n$, there is a least integer r such that $A^r \in \text{Idem}(B_n)$ and a least integer t such that A^t is transitive. Moreover, (i) whenever A^s is transitive $d|s$ where d is the period of oscillation of A, (ii) A^r is the only transitive element in the group of Gr (A) (see Definition 5.4.1), (iii) we always have $t \geq r/n$, and (iv) if $t < r$, then $A^r \leq A^t \odot A^{t+d} \odot \ldots \odot A^{r-d}$.

Proof. We prove (ii) first. If $A^r \in$ Gr (A), and is transitive there is some m such that $(A^r)^m = A^r$. We have $A^r \geq A^{2r} \geq \ldots \geq A^{mr}$ by transitivity, so all these inequalities must be equalities. So $A^r \in \text{Idem}(B_n)$. But Gr (A) being a group has a unique idempotent.

(i) If A^s is transitive, so is A^{sm} for every m. For any sufficiently large m, A^{sm} will be in Gr (A) so it will be idempotent. Then from Theorem 5.2.2, $d|sm$ for any sufficiently large m. Thus $d|s$.

(iii) For any vector v we have $vA^t \geq vA^{2t} \geq \ldots \geq vA^{nt} \geq vA^{(n+1)t}$. But any strictly ascending chain of nonzero vectors in V_n has at most n terms. Thus some inequality must be an equality, and so the last inequality must be an equality. This implies $vA^{nt} = vA^{2nt}$ for all v. So $A^{nt} \in \text{Idem}(B_n)$. Thus $r \leq nt$.

(iv) Since $A^t \geq A^{2t} \geq \ldots \geq A^{mt}$, for any sufficiently large m, $A^{mt} = A^r$ by (ii). Thus $A^t \geq A^r$. Also $A^{t+d} \geq A^{r+d} = A^r A^d = A^r$ and so on. This completes the proof. □

PROPOSITION 5.4.5 [345]. For any $A \in B_n$, $g + g^2 + \ldots + g^d = (A + A^2 + \ldots + A^n)^n$, where g is as in Definition 5.4.1.

Proof. Exercise. □

Next the behavior of the series of powers of Boolean matrix A is considered row by row. The series e_iA, e_iA^2, ... is the series of ith rows of the matrices A, A^2, ... (Recall that e_i is the n-tuple with 1 in the ith component, 0 elsewhere).

NOTATION 5.4.2. Let k_i be the least integer such that vector $e_iA^{k_i}$ is equal to infinitely may others in this series, and let d_i be the least positive integer such that $e_iA^{k_i+d_i} = e_iA^{k_i}$.

PROPOSITION 5.4.7 [345]. d is the least least common multiple of the d_i's.

Proof. We have $k \geq k_i$ for each i by taking at $A^k = A^{k+d}$ row-wise. Further $e_iA^{p+q} = e_iA^p$ holds if and only if $p \geq k_i$, and $d_i|q$. Thus d_i must divide d. But looking row-wise we can observe that for p = max k_i, q = l.c.m. d_i the equation $A^{p+q} = A^p$ holds, where l.c.m. denotes the least common multiple. □

PROPOSITION 5.4.8 [345]. Suppose $A \in B_n$ and the (i, i)-entry of $A + A^2 + \ldots + A^n$ is 1. Then d_i is the greatest common divisor of the integers m for which the (i, i)-entry of A^m is 1.

Proof. Let p be such that $a_{ii}^{(p)} = 1$. Then for any sufficiently large q, $e_iA^pA^q \geq e_iA^q$. Thus $e_iA^q \leq e_iA^{p+q} \leq \ldots \leq e_iA^{d_ip+q}$ for large q. But by the definition of d_i, all these inequalities must be equalities. Thus $e_iA^q = e_iA^{p+q}$ for all sufficiently large q. Thus $d_i|p$.

Yet also, if p is such an integer and is sufficiently large, so is $p + d_i$. Thus d_i is a multiple of the greatest common divisor of all such integers. □

THEOREM 5.4.9 [345]. If $A \in B_n$ and the (i, i)-entry of $A + A^2 + \ldots + A^n$ is nonzero, then $k_i \leq (n - 1)^2 + 1$ and $d_i \leq n$.

Proof. The first assertion follows from Proposition 5.7.8. Let s be the weight of e_iA, and let h be the least positive integer such that $e_i \le e_iA^h$.

Case 1. Let h = n and s = 1. Let

$$v(m) = \sum_{j=1}^{m} e_iA^j, \ m = 1, \ldots, n$$

Then $v(1) \le \ldots \le v(n)$. Suppose for some $m \le n - 1$, $v(m) = v(m + 1)$. Then

$$e_iA^{m+1} \le \sum_{j=1}^{m} e_iA^j, \ e_iA^{m+2} \le \sum_{j=2}^{m} e_iA^j + e_iA^{m+1} \le \sum_{j=1}^{m} e_iA^j$$

In this way $v(m + t) = v(m)$ for all $t > 0$. So $e_i \le e_iA^m$. So $h \le n - 1$. This contradicts the hypothesis of our case and shows $v(m) < v(m + 1)$. Moreover $v(m + 1)$ cannot have more than one 1 entry which $v(m)$ does not have, or a strictly increasing chain $v(1) < \ldots < v(n)$ would be impossible because we would run out of places to add ones too soon.

The hypothesis of our case gives $e_i \le e_iA^n$. Thus if e_iA^n had only one 1, $e_i = e_iA^n$. Thus $k_i = 1$ and the conclusion of the theorem would be true.

Therefore we may assume e_iA^n has weight larger than 1. Let p be the least integer such that e_iA^p has weight larger than 1. By the hypothesis of our case on s, $p \ge 2$. Since $v(p)$ contains exactly one 1 that $v(p - 1)$ does not have, any other 1 entry must be contained in some e_iA^{p-t} for $0 < t < p$. Since p was minimal, e_iA^{p-t} has only this 1 entry and $e_iA^{p-t} \le e_iA^p$. Thus $e_iA^{p-t} \le e_iA^p \le \ldots \le e_iA^{p+(n-1)t}$. This chain has length n + 1 so not all inequalities can be strict. So $e_iA^{p+qt} = e_iA^{p+ct}$ for $q \le n - 2$. So $k_i \le p + qt \le n + (n - 2)(n - 1)$.

Case 2. Either $s > 1$ or $h < n$. In the series $e_iA \le e_iA^{h+1} \le \ldots \le e_iA^{(n-s+1)h+1}$ there are $n - s + 2$ members and the first vector has weight s. So not all inequalities can be strict, and thus

$$e_iA^{qh+1} \le e_iA^{ch+1}, \ q \le n - s$$

Then

$$k_i \le qh + 1 \le (n - s)h + 1 \le (n - 1)^2 + 1$$

This proves the theorem. □

PROPOSITION 5.4.10 [345]. If $A \in B_n$ and the (i, i)-entry of $A + A^2 + \dots + A^n$ is 0, then $k_i \le (n - 2)^2 + 2$.

Proof. Let S be the set of 1 entries in $e_i A$. Let A' be the $(n - 1) \times (n - 1)$ matrix obtained from A by deleting the ith row and column. Then for $j \in S$, A will act on e_j in the same way that A' acts on the corresponding vector. Thus by induction we may assume $k_j \le (n - 2)^2 + 1$. However it follows from the definition of S such that $k_i \le \max_{j \in S} k_j + 1$. □

Theorems similar to the following have been obtained by several authors, see D. Rosenblatt [321], A. L. Dulmage and N. S. Mendelsohn [96], and G. Markowsky [242].

THEOREM 5.4.11 [345]. For any matrix $A \in B_n$, $k \le (n - 1)^2 + 1$ where k is the least positive integer such that $A^{k+p} = A^k$ for some positive integer p.

Proof. This follows from Proposition 5.4.7, Theorem 5.4.9, and Proposition 5.4.10. □

The next result is due to K. H. Kim and F. W. Roush [199] and is slightly better than the preceding result for some matrices.

THEOREM 5.4.12 [199]. Let A, B, $C \in B_n$ and A = BC. Let P be a basis for R(C), regarded as a poset. Let $|P| = r$, s be the largest possible size of an antichain in P. Then the index of A is less than or equal to $(r - 1)s + 1$.

Proof. For a vector v, let k(v) be the least positive integer such that $vA^{k(v)+p(v)} = vA^{k(v)}$ for some positive integer p(v). Then $AA^{k+p} = AA^k$ if $k = \max\{k(v): v \in P\}$ and $p = \Pi\, p(v)$ such that $v \in P$. This is because the rows of A are contained in R(C). Thus the index of A is less than or equal to $\max\{k(v): v \in P\} + 1$.

Next we represent A by a graph whose vertices are the elements of P, with a directed edge from x to y if and only if $xA \geq y$. Thus xA equals the sum of all vertices which are connected to x by a directed edge leading out of x. Likewise xA^i is the sum of all vertices which can be reached from x by directed edge sequences of length i.

Let $x \in P$. Case 1. Some edge sequence exists from x to some vertex greater than or equal to x. Let S be such an edge sequence of shortest possible length. Suppose the length of S is $a \leq s$, $x \leq xA^a \leq \ldots \leq xA^{ar}$. This chain has r + 1 members, yet if a basis for R(C) has r members no chain of nonzero vectors in R(C) which is strictly increasing can have more than r members.

Thus some of the inequalities are equalities. This implies the last inequalities is an equality, and $xA^{a(r-1)} = xA^{ar}$. So $k(x) \leq a(r - 1) \leq s(r - 1)$.

So we may assume the length of S is larger than s. Then by definition there are v, w in S, both occurring before the (s + 1)st member of S, with $v \leq w$. We must have $v < w$, and v occurs before w, else S could be shortened. Also $v \neq x$ or w would be the last member of S and the length of S would not be larger than s. Let $vA^b \geq w$, $v \leq vA^b \leq \ldots \leq vA^{br}$. Again we must have some of these inequalities being equalities, and $vA^{b(r-1)} = vA^{br}$ in particular.

Let $xA^c \geq v$, $vA^h \geq x$. Then

$$xA^{c+b(r-1)} \geq vA^{b(r-1)} \tag{5.4.1}$$

$$vA^{b(r-1)+c+h} \geq xA^{b(r-1)+c} \tag{5.4.2}$$

From these inequalities, $vA^{b(r-1)+c+h} \geq vA^{b(r-1)}$. Suppose this inequality is strict. Then also $vA^{b(r-1)+i(c+h)} > vA^{b(r-1)}$ for any positive integer i. Yet if i is a multiple of b, $vA^{b(r-1)+i(c+h)} = vA^{b(r-1)}$, a contradiction. Thus $vA^{b(r-1)+c+h} = vA^{b(r-1)}$. Therefore by (5.4.1) and (5.4.2), $xA^{c+b(r-1)} = vA^{b(r-1)}$. Thus $xA^{c+b(r-1)+b} = xA^{c+b(r-1)}$. Thus $k(x) \leq c + b(r - 1) \leq s - 1 + (s - 1)(r - 1) \leq (r - 1)s$.

Case 2. No edge sequence from x to a vertex greater than or equal to x exists.

Let x_i denote the points such that (1) there exists a path from x to x_i and for any member y of the path before x_i and no $j > 0$ is $yA^j \geq y$; (2) for x_i there is a $j > 0$ such that $xA^j \geq x_i$. Let t denote the length of such a path. From the argument of the previous case, $k(x_i) \leq (r - t - 1)s$ since we may omit considering the points y on the paths before x_i. Also for $q \geq r$, $xA^q = \Sigma\, x_i A^{q-t}$. Thus $k(x) \leq \max\,\{t + k(x_i), r\}$. Thus unless $r = 1$ or $s = 1$, $k(x) \leq (r - 1)s$. If $r = 1$ or $s = 1$ the theorem can be separately established. □

The following example shows that the above bound is not unreasonable.

EXAMPLE 5.4.1. For $s|r$ we take the poset $Q \times T$ where Q is regarded as a poset with no strict inequalities, and T is the linearly ordered set $\{1, 2, \ldots, r/s\}$. Define a monotone relation on this poset by $(x, y)\ R\ (x + 1, z)$ if $z \leq y$, $(0, y)\ R\ (1, y + 1)$, and $(0, r/s)\ R\ (2, 1)$. Then the index of R is $(rs - 2s + 1)$.

REMARK 5.4.2. One might think that the bound of Theorem 5.4.12 could be replaced by $(rs - 2s + 1)$ in general. However at least for $s = 1$ and $s = r - 1$ this bound is the best possible one.

REMARK 5.4.3. Let R^+ be the set of all nonnegative real numbers. As usual, by means of the homomorphism of semiring $R^+ \to \{0, 1\}$ sending x to 1 if and only if $x \neq 0$, this result gives information about zero patterns of nonnegative matrices.

NOTATION 5.4.3. Let $A \in B_n$. Let s(A) will denote the set of integers i from 1 to n such that some entry of A_{i*} is equal to 1 or some entry of A_{*i} is equal to 1.

The following results are due to S. Schwarz [345].

DEFINITION 5.4.4. A matrix $A \in B_n$ is called a *projection-reducible matrix* if and only if $s(A)$ is disjoint union of two sets S_1, S_2 such that $a_{ij} = 0$ whenever $(i, j) \in S_1 \times S_2$. Otherwise A is called a *projection-irreducible matrix*.

Schwarz's definitions of irreducibility and primitivity are unusual in that they restrict attention to the set of i, j such that there is some nonzero entry in the ith or jth row or column.

EXAMPLE 5.4.2. Let

$$A = \begin{bmatrix} 1 & 0 \\ 0 & 0 \end{bmatrix}$$

Then A is projection-irreducible.

PROPOSITION 5.4.13. A matrix $A \times B_n$ is projection-irreducible if and only if the set of i, j such that (i, j)-entry of $A + A^2 + \dots + A^n$ is 1 is the set $s(A) \times s(A)$.

Proof. Exercise.

DEFINITION 5.4.5. Two Boolean matrices will be called *disjoint* if for no i, j is the (i, j)-entry of both equal to 1.

PROPOSITION 5.4.14. Any Boolean matrix A can be written as a sum of disjoint matrices $A_1 + A_2 + \dots + A_m + B$ where A_i are maximal projection-irreducible matrices less than or equal to A and B is either a zero matrix or a matrix which is not greater than or equal to any projection irreducible matrix. Moreover this decomposition is unique up to the order of summands.

Proof. Exercise.

REMARK 5.4.3. Note that this corresponds to writing the Boolean matrix in a block triangular form

$$\begin{bmatrix} 0 & 0 & 0 & 0 \\ * & A_1 & 0 & 0 \\ * & * & A_2 & 0 \\ * & * & * & 0 \end{bmatrix}$$

where B = A in the region marked with asterisks and is 0 elsewhere.

PROPOSITION 5.4.15. A matrix $A \in B_n$ is projection-irreducible if and only if the set {i, j} such that the (i, j)-entry of $g + g^2 + \dots + g^d$ is nonzero is the set $s(A) \times s(A)$.

Proof. Exercise. □

PROPOSITION 5.4.16. If $A \in B_n$ is projection-irreducible, the $g, g^2, \dots, g^d$ are pairwise disjoint.

Proof. Exercise. □

PROPOSITION 5.4.17. If $A \in B_n$ is projection-irreducible, any d consecutive powers $A^i, A^{i+1}, \dots, A^{i+d-1}$ are pairwise disjoint.

Proof. Exercise. □

PROPOSITIOON 5.4.18. If $A \in B_n$ is projection-irreducible, d is the g.c.d. of the integers p such that A^p has a nonzero mian diagonal entry.

Proof. Exercise. □

PROPOSITION 5.4.19. If $A \in B_n$ is projection-irreducible and t is the least integer such that the (i, i)-entry of A^t is 1, then d is the g.c.d. of the t's.

Proof. Exercise. □

D. Rosenblatt 321 has proved results similar to Propositions 5.4.18, 5.4.19 in the context of graph theory. See Section 5.3.

PROPOSITION 5.4.20 [345]. For any symmetric Boolean matrix A, $d \leq 2$.

Proof. Exercise. □

PROPOSITION 5.4.21 [345]. If $A \in B_n$ is symmetric and projection-irreducible and $n \geq 3$, then (i) if $d = 1$, $k \leq 2n - 2$, and (ii) if $d = 2$, $k \leq n - 2$.

Proof. Exercise. □

PROPOSITION 5.4.22 [345]. If $A \in B_n$ is symmetric and projection-irreducible, then (i) if $n > 3$ and $d = 1$ then $k \leq 2n - 4$, (ii) if $n > 3$ and $d = 1$ then $k \leq 2n - 6$, (iii) if $n = 3$ then $k \leq 2$, and (iv) if $n = 2$ then $k = 1$.

Proof. Exercise. □

N. J. Pullman [291] considers the convergence properties of a series $\{A_1\}$, $\{A_1A_2\}$, $\{A_1A_2A_3\}$, ... of matrices for any series $\{A_1, A_2, \ldots, A_m, \ldots\}$ of matrices. These partial products play the role in the study of non-stationary Markov chains that the powers play in the study of stationary chains. (Suppose the (i, j)-entry in A_m is 1 if and only if a transition from state j at time m - 1 to state i at time m is possible. Then the (i, j)-entry in $A_1A_2 \ldots A_m$ is 1 if and only if a transition from state j at time 0 to state i at time m is possible).

PROPOSITION 5.4.23 [291]. If $A_1, \ldots, A_m \in B_n$, then there exists an index t and columns $c_1, \ldots, c_s$ such that for all $m \geq t$: (i) each c_i is a column of $A_1A_2 \ldots A_m$, (ii) each column of $A_1A_2 \ldots A_m$ is a sum of a subset of $\{c_1, \ldots, c_s\}$.

Proof. The $C(A_1A_2 \ldots A_m)$ satisfy $C(A_1) \supset C(A_1A_2) \supset C(A_1A_2A_3) \supset \ldots$. This chain must have a minimal element $C(A_1A_2 \ldots A_t)$. Then $c_1, \ldots, c_s$ be a column basis for $C(A_1A_2 \ldots A_t)$. This proves the proposition. □

Powers being a special case of products of commuting matrices, we might expect to strengthen the conclusion of the above proposition if we assume further that the A_k commute pairwise.

THEOREM 5.4.24 [291]. If $\{A_1, A_2, \ldots, A_m, \ldots\}$ is a sequence of pairwise commuting $n \times n$ Boolean matrices then there exists a sequence of $s \times s$ permutation matrices $\{Q_1, Q_2, \ldots\}$ and an index t such that for all $m \geq t$, $A_1A_2 \ldots A_t \ldots A_m = BQ_{m-t} \ldots Q_2Q_1C$ for some $n \times s$ matrix B and some $s \times n$ matrix C (independent of m). Furthermore, $C = [I \quad D]T$ where $T \in P_n$.

Proof. According to Proposition 5.4.23 there exists an $n \times s$ matrix B, permutation matrices T_m and an index t such that $A_1A_2 \ldots A_t \ldots A_m = B[I \quad D_m]T_m$. Let $S_m = A_1A_2 \ldots A_t \ldots A_m$, $C = [I \quad D]T$, where $D = D_t$ and $T = T_t$. We first show that each A_m permutes the columns of B when $m > t$.

The columns of B are a column basis for $A_1A_2 \ldots A_{m-1} = S_{m-1}$. Since the A_i commute $S_{m-1}A_m = A_mS_{m-1} = S_m$. Thus A_m applied to the columns of S_{m-1} contains a column basis for S_m. It follows that A_m applied to the column basis vectors of S_{m-1} contains a column basis for S_m. Thus A_m applied to the columns of B gives a set of vectors containing the columns of B. Since the columns of B are distinct, multiplication by A_m must simply permute them.

Consequently $A_m B = BF_m$ for some $s \times s$ permutation matrix F. So putting $Q_k = F_{t+k}$ for each $k \geq 1$, we have

$$S_{t+1} = A_{t+1}S_t = A_{t+1}BC = BQ_1C$$

$$S_{t+2} = A_{t+2}S_{t+1} = A_{t+2}BQ_1C = BQ_2Q_1C$$

$$\cdots\cdots\cdots\cdots\cdots\cdots$$

$$S_m = A_m S_{m-1} = A_m BQ_{m-t-1} \cdots Q_2Q_1C = BQ_{m-t} \cdots Q_2Q_1C$$

This completes the proof. □

N. J. Pullman [291] also investigated properties of infinite products of matrices. These infinite products of matrices occur in the study of inhomogeneous Markov chains.

The following theorem summarizes the most important results about asymptotic forms of Boolean matrices.

THEOREM 5.4.25. Let A be a Boolean matrix and G_A its graph.

(1) A is nilpotent if and only if G_A contains no cycles if and only if there exists a permutation matrix P such that PAP^T has no 1 entries on or above the main diagonal.

(2) A converges to J if and only if G_A is strongly connected and the g.c.d. of the lengths of all cycles in G_A is 1.

(3) The period of A is the least common multiple of the periods of the matrices corresponding to the strong components of G_A.

(4) The period of a strongly connected matrix is the g.c.d. of the lengths of all cycles in its graph. A strongly connected matrix with period p can be put into a block form

$$\begin{bmatrix} 0 & A_1 & 0 & \cdots & 0 \\ 0 & 0 & A_2 & \cdots & 0 \\ 0 & 0 & 0 & \cdots & 0 \\ & & \cdots\cdots & & \\ A_d & 0 & 0 & \cdots & 0 \end{bmatrix}$$

(5) A is indecomposable if and only if G_A is strongly connected.

(6) A periodic strongly connected matrix can be put into the above form where each A_i consists entirely of ones.

(7) Any matrix can be put into a block form such that the main diagonal blocks are either zero or are indecomposable matrices and blocks above the main diagonal are zero.

(8) Let A be a periodic matrix put into the form of (7) and let A_{ij} be a block of A such that one of the main diagonal blocks A_{ii} or A_{jj} is nonzero. Then the (i, j)-block of A^n is $(A_{ii})^{n-1}A_{ij}$ or $(A_{ij}) \cdot (A_{jj})^{n-1}$, respectively.

(9) The index of A is at most $(n - 1)^2 + 1$, if A is an $n \times n$ matrix.

By means of the following algorithm one can find the period of oscillation of an arbitrary large matrix $A \in B_n$ is $O(n^3(\log n))$ time.

ALGORITHM 5.4.1. Let p be a prime number which is at least $(n - 1)^2 + 1$. Expand p in binary notation. Compute the required 2^r powers of A by iterated squaring, and multiply them together to obtain 2^p. The period of A equals the period of A^p, and A^p is periodic.

Determine the strong components of the graph of A^p as follows. Obtain a power B of $I + A^p$ larger than the nth by iterated squaring. Then $B \odot B^T$ is the matrix of the equivalence relation of strong connectedness, where $\odot$ denotes elementwise product.

Then obtain the submatrices A_{kk} of A^p corresponding to the strong components of the graph of A^p. The period of A_{kk} is the number of distinct vectors among the rows of A_{kk}. Since the rows of A_{kk} are either equal or have non-overlapping sets of 1 entries, this number can readily be obtained. Then the period of A is the least common multiple of the period of the A_{kk}.

5.5 Applications

The objective of this section is to point out various ways in which Boolean matrices can be used to formulate and solve problems. Thus we shall present a variety of applications to alert the reader to the art of problem solving in science and engineering by using Boolean matrices.

In the first application here, on a small group of people some relationship is observed such as friendship, acquaintance, spending time together, influence, or respect. This relationship gives a binary relation on the set of people, that is, all ordered pairs of people who have the relationship to each other. From the matrix of this binary relation a researcher might want to find significant subgroups of the sets of people. One way is to study powers of the Boolean matrix (or matrix over $\mathcal{R}$) representing the binary relation.

R. D. Luce and A. D. Perry [236] consider the following sociological problem. *In a group of n people, for some relationship between pairs of people in this group, such as person i communicates with person j, we can form an* $n \times n$ *matrix of 0's and 1's. A 1 is placed in (i, j)-entry if and only if person i has this relationship with person j.*

One of the most significant features of such a group is the existence of cliques.

DEFINITION 5.5.1. A *clique* is a maximal subset of the group of people with the property that person i has the relationship to person j for all $i \neq j$ in the group.

NOTATION 5.5.1. It is assumed no person has the relation to himself, so the main diagonal entries are 0. Let S be the matrix such that $s_{ij} = 1$ if and only if both $a_{ij} = 1$ and $a_{ji} = 1$ where A is the matrix of relation. Then S is the submatrix of symmetric relationships of the group.

PROPOSITION 5.5.1. Person i belongs to some clique if and only if the ith main diagonal entry of S^3 is nonzero.

For the next two theorems we take S^3 to be the cube of S regarded as a nonnegative matrix of real numbers, rather than a Boolean matrix.

THEOREM 5.5.2 [236]. If in S^3, t entries of the main diagonal have the value (t - 2)(t - 1) and all other main diagonal entries are 0, then these t members form a clique, which is the only clique. The converse is also true.

THEOREM 5.5.3 [236]. If an element i is contained in m cliques of which the kth has t_k members, and there are h_k unordered pairs {a, b} common to the kth clique and all the preceding ones, such that i ≠ a, i ≠ b, then

$$s_{ii}^{(3)} = \sum_{k=1}^{m} \left((t_k - 1)(t_k - 2) - 2h_k\right)$$

Proof. The numbers $s_{ii}^{(3)}$ is twice the number of three element sets {i, a, b} contained in some clique. The number of such sets in the kth clique is

$$\frac{(t_k - 1)(t_k - 2)}{2}$$

Such a set is in a preceding one of these cliques if and only if the pair {a, b} is. Thus the number of sets whose first occurrence is in the kth clique is

$$\frac{(t_k - 1)(t_k - 2)}{2} - h_k$$

Thus the total number of such sets, times 2 is given by the formula of the theorem. □

Many researchers have studied these ideas by means of graph theoretical methods: two examples are F. Harary and F. Roberts. H. White has been using different methods to study this subject.

EXAMPLE 5.5.1. If A is

$$\begin{array}{c} \\ 1 \\ 2 \\ 3 \\ 4 \\ 5 \\ 6 \\ 7 \\ 8 \end{array}\begin{array}{c} \begin{array}{cccccccc} 1 & 2 & 3 & 4 & 5 & 6 & 7 & 8 \end{array} \\ \left[\begin{array}{cccccccc} 0 & 1 & 1 & 0 & 0 & 0 & 1 & 0 \\ 1 & 0 & 1 & 1 & 0 & 0 & 1 & 0 \\ 1 & 1 & 0 & 0 & 0 & 1 & 1 & 0 \\ 0 & 0 & 0 & 0 & 0 & 0 & 0 & 0 \\ 0 & 0 & 1 & 0 & 0 & 0 & 0 & 0 \\ 0 & 0 & 0 & 1 & 0 & 0 & 0 & 1 \\ 1 & 1 & 1 & 0 & 0 & 0 & 0 & 0 \\ 1 & 0 & 0 & 0 & 0 & 1 & 0 & 0 \end{array}\right] \end{array}$$

then S is

$$\begin{bmatrix} 0 & 1 & 1 & 0 & 0 & 0 & 1 & 0 \\ 1 & 0 & 1 & 0 & 0 & 0 & 1 & 0 \\ 1 & 1 & 0 & 0 & 0 & 0 & 1 & 0 \\ 0 & 0 & 0 & 0 & 0 & 0 & 0 & 0 \\ 0 & 0 & 0 & 0 & 0 & 0 & 0 & 0 \\ 0 & 0 & 0 & 0 & 0 & 0 & 0 & 1 \\ 1 & 1 & 1 & 0 & 0 & 0 & 0 & 0 \\ 0 & 0 & 0 & 0 & 0 & 1 & 0 & 0 \end{bmatrix}$$

and S^2 is

$$\begin{bmatrix} 3 & 2 & 2 & 0 & 0 & 0 & 2 & 0 \\ 2 & 3 & 2 & 0 & 0 & 0 & 2 & 0 \\ 2 & 2 & 3 & 0 & 0 & 0 & 2 & 0 \\ 0 & 0 & 0 & 0 & 0 & 0 & 0 & 0 \\ 0 & 0 & 0 & 0 & 0 & 0 & 0 & 0 \\ 0 & 0 & 0 & 0 & 0 & 1 & 0 & 0 \\ 2 & 2 & 2 & 0 & 0 & 0 & 3 & 0 \\ 0 & 0 & 0 & 0 & 0 & 0 & 0 & 1 \end{bmatrix}$$

Here the zero main diagonal entries show that 4, 5 have no symmetric relationships with anyone, and 6, 8 have symmetric relationship with each other only, since their main diagonal entries are 1. Hence:

$$\begin{bmatrix} 6 & 7 & 7 & 0 & 0 & 0 & 7 & 0 \\ 7 & 6 & 7 & 0 & 0 & 0 & 7 & 0 \\ 7 & 7 & 6 & 0 & 0 & 0 & 7 & 0 \\ 0 & 0 & 0 & 0 & 0 & 0 & 0 & 0 \\ 0 & 0 & 0 & 0 & 0 & 0 & 0 & 0 \\ 0 & 0 & 0 & 0 & 0 & 0 & 0 & 1 \\ 7 & 7 & 7 & 0 & 0 & 0 & 6 & 0 \\ 0 & 0 & 0 & 0 & 0 & 1 & 0 & 0 \end{bmatrix}$$

By Theorem 5.5.2, {1, 2, 3, 7} are the unique clique.

REMARK 5.5.1. A clique can be contained in the union of other cliques without being contained in any one of them. For instance cliques on {1, 2, 4}, {2, 3, 5}, {1, 3, 6} will together contain a clique on {1, 2, 3}.

R. D. Luce [233] extends the above results. It is to be expected that in addition to exact cliques, there will be subsets of people which are almost exact cliques and for practical purposes should be counted as exact clique.

Theoretical work has been done on the presence of cliques in arbitrary graphs (or matrices) [99, 105, 230, 405].

DEFINITION 5.5.2. A q-*chain* from i to j is a sequence $i = i_0$, $i_1, \ldots, i_q = j$ such that i_k has the given relationship to i_{k+1} for each k from 0 to j - 1.

DEFINITION 5.5.3. A subset of the group is an n-*clique* if and only if (i) it contains at least three elements, (ii) for any pair of elements i, j in it, $i \neq j$ there is some q-chain from i to j (which may include elements of the group outside the clique) for some $q \leq n$, and (iii) the subset is not contained in a larger subset satisfying (ii).

R. D. Luce [233] expects that only 1-cliques (which are cliques), 2-cliques and 3-cliques will be of practical importance.

To determine the n-cliques of a group represented by a Boolean matrix A compute the sum $A + A^2 + \ldots + A^n$ and delete the main diagonal entries of the result to obtain a matrix B. Then the cliques of B can be determined, and they are the same as the n-cliques of A.

EXAMPLE 5.5.2. Let

$$A = \begin{bmatrix} 0 & 1 & 1 & 0 \\ 1 & 0 & 0 & 0 \\ 1 & 0 & 0 & 1 \\ 0 & 0 & 1 & 1 \end{bmatrix}$$

Then

$$A + A^2 = \begin{bmatrix} 1 & 1 & 1 & 1 \\ 1 & 1 & 1 & 0 \\ 1 & 1 & 1 & 1 \\ 1 & 0 & 1 & 1 \end{bmatrix}$$

$$A + A^2 + A^3 = \begin{bmatrix} 1 & 1 & 1 & 1 \\ 1 & 1 & 1 & 1 \\ 1 & 1 & 1 & 1 \\ 1 & 1 & 1 & 1 \end{bmatrix}$$

There are no 1-cliques. The set {1, 2, 3}, {1, 3, 4} are the 2-cliques since they are the cliques of $A + A^2$. The set {1, 2, 3, 4} is a 3-clique since in $A + A^2 + A^3$ for all i, j the (i, j)-entry is 1.

Other work in the sociology of small groups has continued to emphasize the matrices arising from ties, or relationships among the group members. Sometimes several ties, such as liking, esteem, influence, contact are considered for the same group.

In addition to the study of cliques in these matrices other methods of analysis have been proposed. General methods of finding significant subsets of a set on which a binary relation is defined are included in the subject of cluster analysis.

H. C. White, S. A. Boorman, and R. L. Breiger [398] have proposed the view that when a group is to be divided into subgroups, or blocks, the important feature is which pairs of blocks are such that no individual in the first block has any tie to an individual in the second block.

Groups may be analyzed by the following method. The investigator makes a hypothesis, called a *blockmodel*, about the block structure of the group. This hypothesis consists of a Boolean matrix or set of Boolean matrices, or relatively small size m. Then the question is raised: is there a partition of the group into m blocks such that there exists no tie between any individuals of the i and j blocks if and only if the (i, j)-entry of the blockmodel matrix is 0 ? G. H. Heil has written a computer algorithm for doing this, called *BLOCKER*. Specifically, this program finds all assignments of group members to blocks such that the given blockmodel is obtained.

H. C. White, S. A. Boorman, and R. L. Breiger [398] also used a hierarchical clustering algorithm called *CONCOR*, written by R. L. Breiger. It yielded quite similar results to BLOCKER.

R. L. Breiger [42] gives sociological interpretations of different 2×2 blockmodels. This matrix

$$\begin{bmatrix} 1 & 0 \\ 0 & 0 \end{bmatrix}$$

can be interpreted as a caucus, and

$$\begin{bmatrix} 1 & 0 \\ 0 & 1 \end{bmatrix}$$

as a multiple caucus. The matrices

$$\begin{bmatrix} 1 & 0 \\ 1 & 0 \end{bmatrix}, \begin{bmatrix} 1 & 0 \\ 1 & 1 \end{bmatrix}$$

are termed hierarchy and deference, respectively. The structure

$$\begin{bmatrix} 1 & 1 \\ 1 & 0 \end{bmatrix}$$

is known as center-periphery, and

$$\begin{bmatrix} 1 & 1 \\ 1 & 1 \end{bmatrix}$$

may represent an amorphous structure. Some of these structures were earlier defined by P. H. Rossi [323].

If more than one tie is considered by the same group, one sets up a blockmodel matrix for each tie and attempts to divide the group into subsets in such a way that each blockmodel matrix correctly represents its tie.

S. A. Boorman and H. C. White presented the idea that the interrelationship between two or more ties on the same group is related to the structure of the semigroup generated by the corresponding blockmodel matrices.

We next examine the connection between index and period of a Boolean matrix and automata.

Any machine, for instance a computer, can be represented by an automaton. Moreover automata are useful in mathematical linguistics. Here we show that the index and period of a Boolean matrix has significance in the theory of finite state nondeterministic automata. We also mention an application of Bednarek and Ulam [17] in which Boolean matrices can be used for parallel computation, and list references for additional applications.

R. Mandel [238] showed that any n-state, one symbol, nondeterministic automaton has an equivalent one symbol, deterministic automaton which requires no more than $c_n + (n - 1)^2 + 1$ states, where c_n is the largest order of any cyclic subgroup of the symmetric group on $\underline{n}$. Moreover, it can be shown that this is "essentially" the best one can do in general.

The definitions we present here have been modified so as to enable us to relate the results of Section 5.4 to automata theory as quickly as possible. For more general definitions, see any book on automata, such as J. E. Hopcraft and J. D. Ullman [157]. The following results are due to G. Markowsky [242].

DEFINITION 5.5.4. By an *n-state, one symbol, nondeterministic automaton* N we mean a triple (A, S, T) where $A \in B_n$ and $S, T \subset \underline{n}$ where S is the set of initial states and T is the set of final states.

An integer $j \geq 0$ is said to be *good* for N if $SA^j \cap T \neq \emptyset$. Otherwise, j is said to be *bad* for N.

DEFINITION 5.5.5. By an *n-state, one symbol, deterministic automaton* D is a triple (f, a, T) such that $f: \underline{n} \to \underline{n}$, $a \in \underline{n}$, and $T \subset \underline{n}$ where a is the initial state and T is the set of final states.

An integer $j \geq 0$ is said to be *good* for D if $f^j(a) \in T$. Otherwise, j is said to be *bad* for D.

DEFINITION 5.5.6. Two n-state, one symbol automata are said to be *equivalent* if an integer is good for one automaton if and only if it is good for the other.

The basic question here is, given an n-state, one symbol nondeterministic automaton, what is the smallest number of states for which it is possible to have an equivalent one symbol deterministic automaton ? A number of people have shown that for an alphabet of more than one symbol, we must (in the worst case) use 2^n states to realize an n-state, nondeterministic automaton by an equivalent deterministic automaton [239]. R. Mandel [238] shows that at least c_n states (in the worst case) are necessary in the one symbol case and that $c_n + (n - 1)^2 + 1$ states is always sufficient. We will use Theorem 5.4.11 to obtain the same upper bound. (In addition the fact that no cyclic subgroup of B_n has order larger than some cyclic subgroup of P_n (i.e.,

c_n) is used). The construction we use is a slight modification of a standard automata theory construction.

THEOREM 5.5.4 [242]. Let N = (A, S, T) be an n-state nondeterministic automaton. Let K be the collection of all distinct sets of the form SA^j, $j \geq 0$. Let $f: K \to K$ be defined by $f(SA^j) = SA^{j+1}$. Then $D = (f, a, T')$ is equivalent to N where a = S and $T' = \{M \in K: M \cap T \neq \emptyset\}$. The number of states in D (i.e., $|K|$) is less than or equal to $c_n + (n - 1)^2 + 1$.

Proof. Note that j is good for f if and only if $f^j(a) \in T'$ if and only if $SA^j \in T'$ if and only if $SA^j \cap T \neq \emptyset$. Thus the two automata are equivalent. A bound on the size of K is given by the order of the cyclic subsemigroup generated by A, plus 1. The group part of this subsemigroup is a cyclic subgroup of B_n and has order less than or equal to c_n. By the theorem of Schwarz (see Theorem 5.4.11) the non-group part of this semigroup has order at most $(n - 1)^2$. □

For related work see K. H. Kim and F. W. Roush [188].

A. R. Bednarek and S. M. Ulam [17] give an idea which they expect may extend the scope of parallel computation. Given two binary relations, R, S on a set X. A. Tarski observed that $RS = \pi\{(R \times X) \cap (X \times S)\}$ where π is the projection $X \times X \times X \to X \times X$ sending (x_1, x_2, x_3) to (x_1, x_3). They also present some ideas for design of computers which will use this formula to compute the composition of two binary relations by a parallel method. Using discrete (digital) circuit elements such a computer could be built at present. Moreover, these authors give some ideas on how an analogue system might be built, using some type of wave phenomenon.

One such a computer component were built, it would be possible to rapidly compose any two binary relations, and thus to rapidly perform any operation which can be reduced to composition of binary relations. For instance, an approximate integration can be done as follows. Let g be a function from X to X where $X = \{a, a + \Delta x, a + 2\Delta x, \ldots, b\}$ and let H be the binary relation $x \geq y$. Then the composition fH will represent all points below the graph of f. Display this composition on an optical device and let λ be the luminosity that results. Then λ will be proportional to the number of points in $X \times X$ below the curve, which is proportional to

$$\int_a^b f \, dx$$

It may be possible to construct a useful model of the first order predicate calculus using binary relations.

They also propose a theory for how the human brain may recognize patterns. In this theory the brain stores a model of a pattern, and a method for rapidly obtaining many variations on it. For instance there could be several transformations stored as binary relations, and compositions of these could be applied to the model to produce variations on it. To compare variations with an observed picture they propose several metrics on subsets of a set.

First let $s_1(A, B)$ be a real-valued, symmetric function on subsets of $X \times X$. The three examples considered in most detail are $s_1(A, B) = ||A| - |B||$, the Hausdorff metric, and the Steinhaus metric (the measure of the symmetric difference $A \Delta B$ divided by the sum of the measures of A and B). Assuming X is an evenly spaced set of points on [0, 1]. Divide [0, 1] in two and let Q_1, Q_2, Q_3, Q_4 be the four resulting squares. Then divide [0, 1] in four and let $Q_5, Q_6, \ldots, Q_{20}$ be the 16 resulting squares. And so on. Then let $s_1(A, B)$ be

$$\frac{s_1(Q_1 \cap A, Q_1 \cap B)}{4} + \cdots + \frac{s_1(Q_4 \cap A, Q_4 \cap B)}{4}$$

$$\frac{s_1(Q_5 \cap A, Q_5 \cap B)}{4} + \cdots + \frac{s_1(Q_{20} \cap A, Q_{20} \cap B)}{16} + \cdots$$

For additional applications of Boolean matrices to computer science, see H. Jürgensen and P. Wick [168], A. R. Bednarek and S. M. Ulam [17].

In the following we discuss a model of D. Rosenblatt [322] of information diffusion systems in which each member of a set of entities may either derive one item of information from another item, or communicate the item to some other entity. The set of entities and the set of items of information are specified. There are three binary relations: (i) communication relation ρ, this states which entities may communicate with which others, (ii) derivation relation δ, this specifies which items of information can be deduced from which other items of information by any entity, (iii) the assignment relation τ, this states which entities initially possess which items of information. Therefore $x \rho y$ is read "x communicates with y," $x \delta y$ is read "x primitively implies y" and $x \tau z$ is read "entity x is assigned and holds item z of information." Note that ρ is defined on entities and δ is defined on items of information.

Thus our three binary relations represent the principle that in certain finite closed systems information is produced only by the operations of initial assignment, communication, and derivation. In this model, communication and derivation proceed at the same rate.

DEFINITION 5.5.7. Let $\Omega(i)$ denote $\{(x, y)$: at time i entity x possesses item z of information$\}$. Then $\Omega(0) = \tau$, $\Omega(n) = \rho^T\Omega(n - 1) + \Omega(n - 1)\delta$ for $n > 0$, where T denotes transpose. The relation $\Omega(i)$ is called the *thesaurus relation* of order k.

The second equation in Definition 5.5.7 reflects the laws: (i) if y possesses z at time n - 1 and y communicates to x then x possesses z at time n and (ii) if x possesses z at time n - 1 and z implies w then x possesses w at time n.

Does this system converge ? From the above equations we can prove by induction

$$\Omega(m) = \sum_{i=0}^{m} (\rho^T)^i \tau \delta^{m-i}$$

where a zero power of a relation is interpreted as an identity relation. If ρ, τ are reflexive and transitive then the last equation reduces to $\Omega(k) = \rho^T\tau + \tau\delta + \rho^T\tau\delta = \Omega(2)$ for all $k > 2$, and we have rapid convergence. One method of studying the general case is with a larger Boolean matrix.

NOTATION 5.5.2. Let R, S, T be Boolean matrices for the relations ρ, δ, τ. Let $\{M(k): k \geq 0\}$ be the sequence of Boolean matrices for the sequence of relations $\{\Omega(k): k \geq 0\}$.

Now consider the square Boolean matrix

$$B = \begin{bmatrix} R^T & T \\ 0 & S \end{bmatrix}$$

The upper right-hand block of B^{n+1} will then satisfy $M(0) = T$, $M(n) = R^T M(n - 1) + M(n - 1)S$. Thus $M(n)$ is the Boolean matrix of $\Omega(n)$. Thus if B converges, so will the information system.

We generalize the above model by allowing each entity to have a different process of deriving information from other information, and by having different communication networks for different items of information. This generalization is due to K. H. Kim and F. W. Roush [206].

NOTATION 5.5.3. Let X be the set of entities and let Y be the set of possible items of information such that $|X| = m$ and $|Y| = n$. Then a particular system within the model is specified by giving m distinct $n \times n$ Boolean matrices D_i describing for each item of information which items may be derived from it by entity i, as well as n distinct $m \times m$ Boolean matrices C_j giving for item j which entities a given entity may communicate this item to.

An initial state is described by giving a Boolean vector v with nm components which is 1 in a given component if and only if a particular entity x has item y of information. We can form an $nm \times nm$ Boolean matrix H by $h_{(a,b),(c,d)} = 1$ if and only if either $a = c$ and D_a has a 1 in location (b, d) or b = d and C_b has a 1 in location (a, c). Then the state of the system after k steps is given by the vector vH^k where the operations are matrix and matrix-vector multiplication in β_0. So one way to test for convergence, for any v, is to test the matrix H for convergence. For any $H \in B_r$ there is a corresponding graph with r vertices, and a directed edge from x to y if and only if $h_{xy} = 1$.

DEFINITION 5.5.8. A sequence of edges of the form (s_0, s_1), (s_1, s_2), ... , (s_{t-1}, s_t), (s_t, s_0) where the s's need not be distinct, is called a *closed walk.*

Then a necessary and sufficient condition for any Boolean matrix to converge is that the g.c.d. of the lengths of all closed walks through any given vertex is 1, assuming there is at least one closed walk through this vertex.

THEOREM 5.5.5 [206]. The states of the information system will converge for any initial state, provided that one of the following conditions is satisfied: (i) each D_i converges to a matrix which has every entry on its main diagonal equal to 1, and (ii) each C_j converges, and there is a partial order on the set Y such that for each i, $(D_i)_{ab} = 1$ only if a is greater than or equal to b in the partial order. Here Y is the same as in Notation 5.5.3.

Since there is an exact duality between the entities and the items of information in our model, the duals of these conditions also suffice.

Proof. Suppose (i) is satisfied. Let V be a closed walk in the graph of H. Let i be some entity which is involved in V, and so suppose it is involved with item j of information. Then in the graph of D_i there are closed walks through each point, since the main diagonal entries of large powers will be 1. Moreover by convergence there will be a closed walk through the point j of length relatively prime to that of V. Let U be V with this closed walk inserted from the point (i, j) to itself. Then U and V have relatively prime lengths. This verifies the convergence condition for H.

Suppose (ii) is satisfied. Again consider a closed walk V in the graph of H. Some edges in V come from matrices C_j and others from D_i (entries on the main diagonal of H may come from either source). Since C edges will not change the item of information, and the D edges will it the same or decrease it, for a closed walk to occur, the item of information must remain constant. If any edges of some D are involved, they will correspond to diagonal edges in a D matrix and a closed walk U of relatively prime length can be obtained from V by inserting another copy of such an edge. If no D edges are involved the closed walk is a closed walk of some C_j, and the convergence condition for C_j will imply a closed walk of relatively prime length. This proves the theorem. □

EXAMPLE 5.5.3. The graph of H is

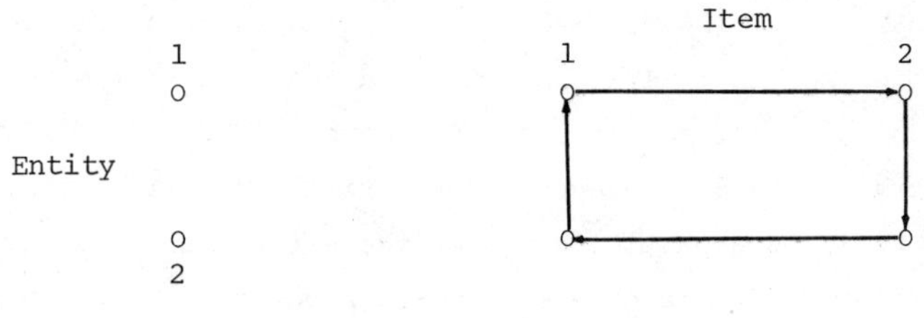

EXAMPLE 5.5.4. All C matrices are zero, some D matrix does not converge.

EXAMPLE 5.5.5. The matrices

$$D_1 = \begin{bmatrix} 1 & 0 & 0 & 0 \\ 1 & 0 & 0 & 0 \\ 1 & 1 & 0 & 0 \\ 1 & 1 & 1 & 1 \end{bmatrix}, \qquad D_2 = \begin{bmatrix} 1 & 0 & 0 & 0 \\ 1 & 0 & 1 & 0 \\ 1 & 0 & 0 & 0 \\ 1 & 1 & 1 & 1 \end{bmatrix}$$

$$C_1 = \begin{bmatrix} 0 & 0 \\ 0 & 0 \end{bmatrix}, \quad C_2 = \begin{bmatrix} 0 & 1 \\ 0 & 0 \end{bmatrix}, \quad C_4 = \begin{bmatrix} 0 & 0 \\ 1 & 0 \end{bmatrix}, \quad C_4 = \begin{bmatrix} 0 & 0 \\ 0 & 0 \end{bmatrix}$$

A. Shimbel [364] has developed a somewhat similar model, except that there is no derivation relation and the initial assignment consists of one item of information to each person. Here too powers of a Boolean matrix are important.

At time 0 each of n persons is given a certain message which the others do not know. At times $t = 1, 2, \ldots$ he communicates all the messages he knows, including those received at preceding times from others, to each member of a subset of the group.

DEFINITION 5.5.9. A Boolean matrix A is called a *communication matrix* if $a_{ij} = 1$ if and only if member i communicates to member j, and $a_{ii} = 1$ for every i.

REMARK 5.5.2. The ith column of the kth power of A tells which messages member i knows at time k.

DEFINITION 5.5.10. The communication matrix A is called *adequate* if and only if eventually all players will know all the messages given to all the other members.

More general systems are also considered in which the communication matrix A is not constant, but varies with time.

Three measures of the efficiency of such as system are given: (i) the reciprocal of the time required for all members to know all the messages, (ii) the number of channels per member is p/n, where p is the number of 1 entries off the main diagonal of a communication matrix A, and (iii) the redundancy ration is n^2/q, where q is the sum of all entries of the kth power of A regarded as a nonnegative matrix, where k is the time required for all members to know all the messages.

EXAMPLE 5.5.6. Let

$$A = \begin{bmatrix} 1 & 1 & 0 & 0 \\ 0 & 1 & 1 & 0 \\ 0 & 0 & 1 & 1 \\ 1 & 0 & 0 & 1 \end{bmatrix}$$

Then p, the number of off-main diagonal 1 entries, is 4 so $p/n = 1$.

It can be seen that A^3 is the least Boolean power of A which has every entry 1, so the reciprocal of the solution is 1/3.

The cube of A regarded as a matrix over $\mathcal{R}$ is

$$\begin{bmatrix} 1 & 3 & 3 & 1 \\ 1 & 1 & 3 & 3 \\ 3 & 1 & 1 & 3 \\ 3 & 3 & 1 & 1 \end{bmatrix}$$

so the redundancy ratio is $16/32 = 1/2$.

For related results, see A. Shimbel [365]. Some uses of Boolean methods in operations research are given in P. L. Hammer and S. Rudeanu [136] and the other papers of P. L. Hammer and his coworkers listed in the bibliography.

F. E. Hohn and L. R. Schissler [156] studied those properties of Boolean matrices which have application to the design of combinatorial relay logic circuits, and developed fundamental aspects of this application.

What will be the effect of a given switching circuit ? We represent a circuit by a graph and the graph by a Boolean matrix and then insert Boolean variables for the relays. The output is represented by an idempotent power of this convergent matrix.

DEFINITION 5.5.11. The *output matrix* is a matrix whose (i, j)-entry tells whether or not there is some direct or indirect electrical path from i to j.

Switching circuits like

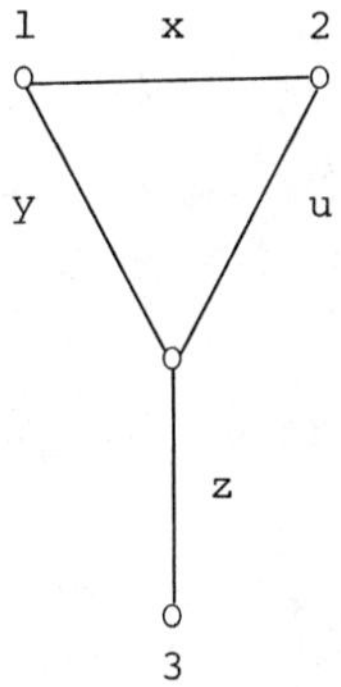

are considered. Here 1, 2, 3 are terminals, and x, y, u are switches or relays which can connect or break the indicated electrical path.

Such circuits are treated by means of Boolean matrices whose entries may involve the variables x, y, u, z in addition to 0's and 1's.

EXAMPLE 5.5.7. For the above circuit this is

$$\begin{bmatrix} 1 & x + yu & xuz + yu \\ x + yu & 1 & xyz + uz \\ xuz + yz & xyz + uz & 1 \end{bmatrix}$$

For instance, current can travel from 1 to 2 either if the x relay is closed or if both the y and the u relays are closed so the (1, 2)-entry is x + yu.

REMARK 5.5.3. Output Boolean matrices are always symmetric, reflexive, and idempotent.

DEFINITION 5.5.12. A *primitive connection matrix* is a matrix which may involve connections other than terminals (of which there is one in the diagram above). For two such connections, or nodes i, j the primitive connection matrix has (i, j)-entry the value of whatever relay lies on the edge from i to j if there is such an edge, and is 0 otherwise.

EXAMPLE 5.5.8. The primitive connection matrix of the circuit above is

$$\begin{bmatrix} 1 & x & 0 & y \\ x & 1 & 0 & u \\ 0 & 0 & 1 & z \\ y & u & z & 1 \end{bmatrix}$$

Primitive connection matrices can thus be written down immediately from the graph of the circuit.

REMARK 5.5.4. Primitive connection matrices are always symmetric and reflexive but not in general idempotent.

Analysis of circuits is concerned with the question: *given a primitive connection matrix, how can the output matrix be obtained* ?

The first step is to remove all nodes that are not terminals (so the node 4 in the matrix above), one at a time. The row and column of such a node (say node r) are deleted and to an (i, j)-entry of the remaining matrix is added the product of the (i, r) and (r, j)-entries in the former matrix.

For the present example this gives

$$\begin{bmatrix} 1 + yy & x + yu & yz \\ x + yu & 1 + uu & uz \\ yz & uz & 1 + zz \end{bmatrix}$$

which equals

$$\begin{bmatrix} 1 & x + yu & yz \\ x + yu & 1 & uz \\ yz & uz & 1 \end{bmatrix}$$

Now take the least power of the matrix resulting which is idempotent. This will be the output matrix.

In the present case the square of the matrix just given is idempotent and is the output matrix.

The problem of greatest practical interest is synthesis of circuits. This may be done by reversing the preceding steps, going from an output matrix to a primitive connection matrix.

Additional results on switching circuits are contained in M. Davio and J.-J. Quisquater [65], P. Delsarte and J.-J. Quisquater [67], M. Davio, J.-P. Deschamps and J.-C. Lienard [64], M. Davio and G. Bioul [62]. See also W. V. Quine [292], E. J. McCluskey [247], A. Varga, P. Ecsedi-Toth and F. Moricz [389], and K. H. Kim and F. W. Roush [206].

F. Robert [316] uses Boolean matrices to study the problem of discrete iterations: given finite sets $X_1, X_2, \ldots, X_n$ and a transformation F on $X_1 \times X_2 \times \ldots \times X_n$, study the powers of F in terms of convergence and fixed points. Problems of this nature arise in information theory, discrete automata, biomathematics, physics, and sociology. Robert's results on Boolean matrices deal mainly with their eigenvalues and eigenvectors. He proves analogues of the Perron-Frobenius and Stein-Rosenberg theorems, characterizing existence of eigenvalues 0 and 1. He then defines Boolean contraction, proves that Boolean contractions have unique fixed points, and develops discrete analogues of Newton's method and other topics of numerical analysis.

B. G. Mirkin [256] employs Boolean matrix methods in the analysis of qualitative data. For instance, attributes can be described by an object-attribute matrix T, such that $t_{ij} = 1$ if object i has attribute j and $t_{ij} = 0$ if object i does not have attribute j. Similarly, Boolean object-object matrices can be used. Methods of analyzing such matrices for the two goals of describing or predicting a new attribute by a combination of old ones, and construction of new attributes which explain experimental results, are given in detail in his book.

Lastly, K. H. Kim and F. W. Roush [206] have applied Boolean matrices to a problem in game theory and clustering analysis.

Exercises

1. Let G be the graph with matrix

$$\begin{bmatrix} 0&0&0&0&0&1&0&0 \\ 0&0&0&0&1&0&0&0 \\ 0&0&1&0&0&0&1&0 \\ 0&1&0&0&0&0&1&0 \\ 0&0&1&1&0&0&0&0 \\ 1&0&0&0&0&0&0&0 \\ 0&0&0&0&0&0&1&1 \\ 0&0&1&0&0&0&0&0 \end{bmatrix}$$

   Determine (a) the reachability matrix of G, (b) the strong components of G, (c) the weak components of G, (d) the matrix of the condensation of G, (e) the connectedness matrix of G, (f) the distance matrix of G, (g) the period of A, (h) the index of A, (i) the least transitive power of A, and (j) the least idempotent power of A.
2. Find all 3×3 nearly decomposable Boolean matrices.
3. Prove every nearly decomposable matrix is prime.
4. Determine all $n \times n$ Boolean matrices of index exactly $(n - 1)^2 + 1$.
5. Count the number of 3×3 and 4×4 Hall matrices.
6. Write an efficient computer program to determine the weak components and strong components of the graph given by a Boolean matrix A, and whether the matrix is primitive.
7. If A is a primitive Hall matrix, show that A^k is fully indecomposable for some $k \in \underline{n}$. What is the least exponent for which this is always true ?
8. Prove Proposition 5.1.2.
9. Prove Theorem 5.2.2.
10. Prove Proposition 5.2.14.
11. Prove Theorem 5.2.20.
12. Prove Proposition 5.4.3.
13. Prove Proposition 5.4.5.
14. Prove Proposition 5.4.13.
15. Prove a reflexive matrix is fully indecomposable if and only if its off-main diagonal entries form an indecomposable matrix [46].

16. Let M be an indecomposable matrix of period $d > 2$. Using the block form of Theorem 5.2.9 and the previous exercise show $I + M$ is fully indecomposable and that no main diagonal entry of $I + M$ lies in a 2×2 rectangle of ones. It follows as in Section 2.4 that $I + M$ is prime [46, 131].

17. Using the preceding exercise for $d = 3$, show there exist at least

$$\frac{n^2}{3} - O(n)$$

prime matrices in B_n.

18. By induction on m show a reflexive matrices in B_{3m} none of whose main diagonal entries lie in a 2×2 rectangle of ones can have weight at most $3m^2 + m$.

19. Show a Boolean matrix is prime if and only if its row space is maximal among row spaces of non-permutation matrices [46].

20. Show that if $A \in B_n$ is fully decomposable such that every rectangle of ones other than a row or column has fewer ones than any row or column, A is prime. It follows that an (n, k, λ)-design is prime if $k > \lambda^2$ [46].

21. Let $A \in B_n$ be nearly decomposable. Show that if $a_{ij} = 1$ then there exist sets S, T such that for $s \in S$, $t \in T$, $a_{st} = 0$ unless $(s, t) = (i, j)$, and $i \in S$, $j \in T$ and $|S| + |T| = n$. (If we change any 1 entry to 0 we have a decomposable matrix).

22. Let $A \in B_n$ be nearly decomposable, and let $a_{i_1 j_1} = a_{i_1 j_2} = 1$. Let S_1, T_1 and S_2, T_2 be as in the preceding exercise. Prove that one of these cases holds:

Case 1. One of $|S_1 \cup S_2|$, $|T_1 \cup T_2|$ is n, the other is $n - 1$, $|S_1 \cap S_2| = 1$, $T_1 \cap T_2 = \emptyset$

Case 2. $|S_1 \cup S_2| = n$, $T_1 \cap T_2 = \emptyset$

Case 3. $S_1 \cap S_2 = \{i_1\}$, $|T_1 \cup T_2| = n$

To do this, consider the rectangles $S_1 \cap S_2$, $T_1 \cup T_2$ and $S_1 \cup S_2$, and $T_1 \cap T_2$.

23. Use the previous exercise to show A has 2×2 rectangles of ones.

24. Use the previous exercise. In similar fashion show a configuration

$$\begin{bmatrix} 1 & 1 & 1 \\ 1 & * & * \\ 1 & * & * \end{bmatrix}$$

cannot exist in A. This proves the first statement of Theorem 5.3.12.

25. Prove the other parts of Theorem 5.3.12 from the preceding exercises.

26. Show there exists a $2n \times 2n$ indecomposable Boolean matrix of period 2 whose index is $2.(n - 1)^2 + 2$. Divide it into 4 blocks, let two blocks be 0, one block an identity matrix, and another a matrix which is primitive of index $(n - 1)^2 + 1$.

27. Generalize the preceding exercise to arbitrary periods d.

28. What is the maximin index of a fully indecomposable matrix ? Give an example.

29. Give a Boolean matrix A such that A^2 has smaller weight than A, but A is primitive. This can be done graphically, or a block matrix (blocks J or 0, of unequal size) can be found and an extra 1 entry added.

30. Suppose $A \times B_n$ is a primitive matrix such that A^s has a main diagonal 1 entry. Bound the index of A in terms of s, n. (This was done by A. L. Dulmage and N. S. Mendelsohn [95]).

31. What is the maximum index of a nilpotent Boolean matrix ? of a reflexive matrix ? of a subtriangular matrix ?

32. Prove that for any $k < \frac{1}{2}$, as $n \to \infty$ the probability tends to 1 that a random $n \times n$ Boolean matrix is kn indecomposable.

APPENDIX

A.1 Matrices Over an Arbitrary Boolean Algebra

Matrices over an arbitrary Boolean algebra β satisfy most of the properties of matrices over $\beta_0 = \{0, 1\}$. The reason is that any Boolean algebra is a sub-Boolean algebra of β_0^S for some set S, and we have an isomorphism from n × n matrices over β_0^S to B_n^S. For example we can define concepts of permutation matrix, inverse, and so on. A permutation matrix is a matrix P such that $PP^T = P^TP = I$. These turn out to correspond to n-tuples of ordinary permutation matrices. If AB = I then A, B are permutation matrices. Any matrix having a g-inverse has a g-inverse which is a partial permutation matrix. There exists a g-inverse if and only if there exists a space decomposition. In fact, G. Markowsky [244] and P. P. Rao [296] have studied this. Infinite dimensional matrices over β_0 do have some different properties (J and D-classes need not coincide). They have been studied by various Russian workers such as K. A. Zaretski. See T. S. Blyth [27, 30] for various results about structures related to infinite Boolean matrices.

A.2 Fuzzy Matrices

L. A. Zadeh [408] has defined a *fuzzy subset* of a set to be a function from the set to [0, 1]. Let S be a nonempty set. This allows us to say, instead of $x \in S$ or $x \notin S$, that x belongs to S with degree 0.9 or some other number. Intersections and unions of fuzzy subsets are defined by infimums and supremums.

We studied the corresponding matrix notion in K. H. Kim and F. W. Roush [193].

DEFINITION A.2.1. A *fuzzy matrix* is a matrix whose entries lie in [0, 1]. Fuzzy matrices are operated upon by the following rules:

$$A + B = \left(\sup \{a_{ij}, b_{ij}\}\right)$$

$$AB = \left(\sup_{k} \{\inf \{a_{ik}, b_{kj}\}\}\right)$$

$$A \odot B = \left(\inf \{a_{ij}, b_{ij}\}\right)$$

$$cA = \left(\inf \{c, a_{ij}\}\right), \quad c \in R$$

EXAMPLE A.2.1. Let

$$A = \begin{bmatrix} 0.8 & 0.1 \\ 0.2 & 1 \end{bmatrix}, \qquad B = \begin{bmatrix} 0.6 & 0.5 \\ 0.4 & 0.3 \end{bmatrix}$$

Then

$$A + B = \begin{bmatrix} 0.8 & 0.5 \\ 0.4 & 1 \end{bmatrix}, \qquad AB = \begin{bmatrix} 0.6 & 0.5 \\ 0.4 & 0.3 \end{bmatrix}$$

These are precisely matrices over the fuzzy algebra [0, 1] under the operations sup {a, b} and inf {a, b}. Fuzzy matrices form a partially ordered semiring, as Boolean matrices do. *Boolean matrices form a subset of the set of fuzzy matrices.* Conversely there exists a homomorphism h_α from the semiring of fuzzy matrices to the semiring of Boolean matrices, where $h_\alpha(A)_{ij} = 0$ if $a_{ij} < \alpha$ and $h_\alpha(A)_{ij} = 1$ if $\alpha_{ij} \geq \alpha$. Two fuzzy matrices are equal if and only if their images are equal under all these homomorphisms.

Fuzzy matrices can be used in studying clustering and in the theory of fuzzy networks. In K. H. Kim and F. W. Roush [193, 195] proved results about rank, basis, D-classes, idempotents, regularity, inverses, eigenvectors, index, and period for fuzzy matrices. The most important difference is that bases (minimal spanning sets) are not unique.

DEFINITION A.2.2. A basis C over the fuzzy algebra is a *standard basis* provided that if $C<i> = \sum a_{ij}C<j>$ for $C<i>, C<j> \in C$, then $a_{ii}C<i> = C<i>$.

EXAMPLE A.2.2. The basis {(0.5, 1, 0.5), (0, 1, 0.5)} is not standard since (0.5, 1, 0.5) = (0.5)(0.5, 1, 0.5) + (0, 1, 0.5) but {(0.5, 0.5, 0.5), (0, 1, 0.5)} is a standard basis for the same subset.

Lastly, we remark that there exists a unique standard basis for every finitely generated subspace of fuzzy vectors.

OPEN PROBLEMS

1. What is the distribution of the row rank, column rank, and Schein rank of a random Boolean matrix ?
2. (P. Erdös and L. Moser [101]). Let $n_1(V_n)$ be the least number of elements of V_n such that every vector of V_n is a sum of two of the $n_1(V_n)$ vectors. Is $n_1(V_{2n}) > 2^n(1.75)$ or $n_1(V_{2n}) < 2^n(1.75)$ for large n ?
3. (P. Erdös and L. Moser [101]). Let $n_1(V_n)$ be as in Problem 2. Let $n_2(V_n)$ be the largest size of a set of Boolean vectors v_i, $i = 1$ to $n_1(V_n)$, such that all the vectors $v_i + v_j$ for $1 \le i < j \le n$ are distinct. Find c_1, c_2 with $0 < c_1 < c_2 < 1$ and $c_2/c_1 < 1.01$ and $(1 + c_1)^n < n_1(V_n) < (1 + c_2)^n$, for large n.
4. Find the least number of transformations and transpose transformations defined on $\underline{n}$ whose sum is greater than the matrix of a linear order. (An upper bound is $[\frac{2n+2}{3}]$).
5. Is shift equivalence decidable if the characteristic polynomials of two matrices A, B over R are allowed to have multiple roots ?
6. What is the number of posets on $\underline{n}$ for $n \ge 12$?
7. What is the number of quasi-ordered sets on $\underline{n}$ for $n \ge 12$?
8. Let p_n be the number of posets on $\underline{n}$. Show that the number of isomorphism classes of topologies on $\underline{n}$ is asymptotically equal to $p_n/n!$.
9. What is the asymptotic number of Idem (B_n) ?
10. What is the cardinality of Idem (B_n) ?
11. What is the asymptotic number of Reg (B_n) ?
12. What is the cardinality of Reg (B_n) ?

13. What is the number of Hall matrices in B_n ?
14. What is the number of transitive relations in B_n ?
15. What is the number of indecomposable matrices in B_n ?
16. Find a formula for the number of similarity relations and of connective relations on a product of two linear orders.
17. Find a complete description of Pmt (B_n) or some related semigroup. (For instance the semigroup generated by all primitive matrices and all permutations). (See Section 5.3).
18. Find the asymptotic number of nearly decomposable matrices in B_n.
19. Find the number nearly decomposable matrices in B_n.
20. What is the maximum permanent of an $n \times n$ nearly decomposable Boolean matrix ?
21. The *complication* of a Boolean function is the least number of operations $\wedge$, $\vee$, required to express it (parentheses and complements may also be used but are not counted). Can any Boolean function be proved to have large complications ? For instance does the function which equals 1 if and only if the number of x_i which equal 1 is exactly $\frac{n}{2}$ have complication larger than any polynomial function ? Perhaps the function which equals 1 if and only if $x_1, \ldots, x_n$ represents a prime number in binary notation would have complication close to the maximum.
22. (P. L. Hammer and S. Rudeanu [136]). Let G be a graph with vertices v_i which are to be 4-colored. Let $a_{ij} = 1$ or $a_{ij} = 0$ according as v_i is or is not connected to v_j by an edge. Let x_i, y_i be Boolean variables such that $x_i = 1$ if and only if v_i is colored with colors 1 or 2, $y_i = 1$ if and only if v_i is colored with colors 1 or 3. Then a Boolean equation expressing the condition that G can be 4-colored is

$$\sum_{i,j} a_{ij}(x_i y_i x_j y_j + x_i y_i^c x_j y_j^c + x_i^c y_i x_j^c y_j + x_i^c y_i^c x_j^c y_j^c) = 0$$

where x^c denotes the complement of x. Find a Boolean proof of the 4-color theorem.

23. Show that for any finite Boolean algebra and any finite or infinite collection of Boolean matrices A_α the sum and the elementwise product of all the A_α are well-defined. Are there any cases in which the matrix product of all the A_α is well-defined ? Perhaps one could base this on some sort of convergence of finite subproducts.
24. What is the height of the partial order of D-classes in B_n ? (K. H. Kim and F. W. Roush [189] have solved this for regular D-classes).
25. What is the probability that a random element of B_n, $n \to \infty$ has Schein rank less than n ?
26. What is the probability that a random element of B_n is prime ?
27. Is there a polynomial time algorithm, given a finite set of Boolean matrices, to determine whether universal matrix lies in the semigroup they generate ? (K. H. Kim and F. W. Roush [205] have solved this for Boolean matrices with no zero rows or columns). What about 0 ?
28. The following is a well-known unsolved combinatorial problem. Determine the function f(n) such that every $f(n) \times f(n)$ matrix has some $n \times n$ rectangle of zeros or some $n \times n$ rectangle of ones. (See P. Erdös and J. Spencer [105]).
29. What is the asymptotic number of D-classes of B_{nm} where

$$\frac{n}{\log m} \to 0 \ ?$$

30. Is the problem of finding a square root of a Boolean matrix NP-complete ?

REFERENCES

1. A. Abian, Boolean Rings, Brandon Press, Boston, Mass., 1976.

2. A. J. Aizenstatt, On class of periodic semigroups (Russian), Ucenye Zap. Leningrad Gos. Ped. Inst., A. I. Gercen 183 (1958), 241-249.

3. A. J. Aizenstatt, On defining relations of symmetric semigroups (Russian), Mat. Sbornik 45 (1958), 261-280. MR 21#86.

4. A. J. Aizenstatt, On defining relations of symmetric semigroups (Russian), Ucenye Zap. Leningrad Gos. Ped. Inst., A. I. Gercen 166 (1958), 121-142.

5. A. J. Aizenstatt, On the semigroup of all one-to-one mappings of natural numbers into itself (Russian), Ucenye Zap. Vybork. Gos. Ped. Inst. 2 (1957), 15-24.

6. V. Alekseev, Quasirandom dynamical systems, Mat. Sbornik 76-118 (1968), 72-134.

7. P. S. Alexandrov, Combinatorial Topology, Vol. 1, Graylock, Rochester, New York, 1956.

8. D. Allen, Jr., A generalization of the Rees theorem to a class of regular semigroups, Semigroup Forum 2 (1971), 321-331.

9. S. S. Anderson, Graph Theory and Finite Combinatorics, Markham, Chicago, 1970.

10. S. S. Anderson and G. Chartrand, The lattice-graph of the topology of a transitive directed graph, Math. Scand. 21 (1967), 105-109.

11. G. Andreoli, Groups of substitutions that operate on an infinity of elements (Italian), Giorn. Mat. Battaglini 88 (1960), 127-134.

12. K. J. Arrow, Social Choice and Individual Values, 2nd ed., Yale Univ. Press, New Haven, Conn., 1963.

13. P. C. Baayen, P. Van Emde Boas, and D. Kruyswijk, Combinatorial Problem on Finite Semigroups, Math. Centrum, Amsterdam, 1965.

14. J. Baillieul, Green's relations in finite function semigroups, Aeq. Math. 7 (1971), 22-27.

15. H.-J. Bandelt, Regularity and complete distributivity, Semigroup Forum 19 (1980), 123-126.

16. S. Barbera, Manipulation of social decision functions, Jour. of Econ. Theory 15 (1977), 266-278.

17. A. R. Bednarek and S. M. Ulam, On the theory of relational structures and schemata for parallel computation, Informal Report LA-6134-MS, Los Alamos Scientific Lab., Los Alamos, New Mexico, 1977.

18. A. Ben-Israel and A. Charnes, Contribution to the theory of generalized inverses, Jour. of SIAM Appl. Math. 11 (1963), 667-699.

19. A. Ben-Israel and T. E. N. Greville, Generalized Inverses: Theory and Applications, John Wiley, New York, 1974.

20. C. Berge, Graphs and Hypergraphs, Elsevier North-Holland, Amsterdam, 1973.

21. C. Berge, The Theory of Graphs, John Wiley, New York, 1964.

22. G. Birkhoff, Lattice Theory, Amer. Math. Soc., Providence, R. I., 1967.

23. On groups of automorphisms, Revista Union Mat. Argentina 11 (1946), 155-157.

24. D. Black, On the rationale of group decision making, Jour. of Political Economy 56 (1948), 23-34.

25. D. Black, The Theory of Committees and Elections, Cambridge Univ. Press, Cambridge, England, 1958.

26. R. L. Blair, Stone's topology for binary relations, Duke Math. Jour. 22 (1955), 271-280.

27. T. S. Blyth, $\wedge$-distributive Boolean matrices, Proc. of Glasgow Math. Assoc. 7 (1965), 93-100.

28. T. S. Blyth, Modules and matrices on stack (French), Bull. Soc. Royal Sci. de Liege 39 (1970), 451-469.

29. T. S. Blyth, On eigenvectors of Boolean matrices, Proc. of Royal Soc., Edinburgh Sec. A 67 (1966), 196-204.

30. T. S. Blyth, An open question on similarity of Boolean matrices (French), C. R. Acad. Sci. Paris, 266 (1968), 963-965.

31. G. Boole, An Investigation of the Laws of Thought, on Which are Founded the Mathematical Theories of Logic and Probabilities, Macmillan, Cambridge, 1854.

32. G. Boole, The Mathematical Analysis of Logic, Being an Essay Towards a Calculus of Deductive Reasoning, Macmillan, Cambridge, 1847.

33. J. Borosh, D. J. Hartfiel, and C. J. Maxson, Answers to questions posed by Richman and Schneider, Linear Alg. & Multilinear Alg. 3 (1976), 255-258.

34. J. Bosak, Directed graphs, Preprint, Bratislava, Czechoslavakia, 1977.

35. R. Bowen, Markov partitions for axiom A diffeomorphisms, Amer. Jour. Math. 42 (1970), 725-747.

36. R. Bowen, Topological entropy and axiom A, Proc. Symp. Pure Math. Vol. 14, Amer. Math. Soc., Providence, R. I., 1970, 23-42.

37. R. Bowen and O. Langford, Zeta functions of restrictions of the shift transformation, Proc. Symp. Pure Math. Vol. 14, Amer. Math. Soc., Providence, R. I., 43-50.

38. V. J. Bowman and C. S. Colantoni, The extended Condorcet condition: A necessary and sufficient condition for the transitivity of majority decision, Jour. of Math. Sociology 2 (1972), 267-283.

39. G. H. Bradley, P. L. Hammer, and L. Wolsey, Coefficient reduction for inequalities in 0, 1 variables, Preprint, Univ. of Waterloo, Waterloo, Canada, 1977.

40. R. L. Brandon, D. W. Hardy, K. H. Kim, and G. Markowsy, Cardinalities of D-classes in B_n, Semigroup Forum 4 (1972), 341-344.

41. R. L. Brandon, D. W. Hardy, and G. Markowsky, The Schützenberger group of an H-class in the semigroup of binary relations, Semigroup Forum 5 (1972), 45-53.

42. R. L. Breiger, Network analysis of a community elite, Lecture Delivered at the Annual Meeting of Public Choice Society, New Orleans, March 1977.

43. F. M. Brown, Boolean Matrices and Finite Systems, Thesis, Dept. of Elec. Eng., Ohio State Univ., Columbus, Ohio, 1967.

44. R. H. Bruck, A Survey of Binary Systems, Ergebnisse der Math. Heft 20, Springer-Verlag, Berlin, 1958.

45. N. G. De Bruijn, C. A. Van E. Tengbergen, and D. R. Kruywijk, On the set of divisors of a number, Nierw. Arch. Wisk. 23 (1952), 191-193.

46. D. De Caen and D. A. Gregory, Primes in the semigroup of Boolean matrices, Preprint, Queen's Univ., Toronto, Canada, 1980.

47. A. Cayley, The theory of groups: Graphical representation, Amer. Jour. Math. 1 (1878), 174-176.

48. C.-Y. Chao, On a conjecture of the semigroup of fully indecomposable relations, Preprint, Univ. of Pittsburgh, Pittsburgh, Pa., 1977.

49. C.-Y. Chao, On maximal subgroups of the semigroups of binary relations, Kyungpook Math. Jour. 17 (1977), 143-152.

50. C.-Y. Chao and S. Winograd, A generalization of a theorem of Boolean relation matrices, Preprint, Univ. of Pittsburgh, Pittsburgh, Pa., 1975.

51. S. D. Chatterji, The number of topologies on n points, NASA Report, Kent State Univ., Kent, Ohio, 1966.

52. R. Chaudhurd and A. Mukherjea, Idempotent Boolean matrices, Semigroup Forum 21 (1980), 273-282.

53. V. Chvatal and P. L. Hammer, Set-packing problems and threshold graphs, Preprint, Univ. of Waterloo, Waterloo, Canada, 1976.

54. A. H. Clifford, A proof of the Montague-Plemmons-Schein theorem on maximal subgroups of the semigroup of binary relations, Semigroup Forum 1 (1970), 272-275.

55. A. H. Clifford, Semigroups admitting relative inverses, Ann. Math. 42 (1941), 1037-1049.

56. A. H. Clifford and G. B. Preston, Algebraic Theory of Semigroups, Vol. 1, Amer. Math. Soc., Providence, R. I., 1961.

57. E. F. Codd, Cellular Automata, Academic Press, New York, 1968.

58. L. Comtet, Coverings, filter bases, and topologies on a finite set (French), C. R. Acad. Sci. Paris, Ser. A-B 262 (1966), A1091-A1094.

59. I. M. Copilowish, Matrix development of the calculus of relations, Jour. of Symbolic Logic 13 (1948), 193-203.

60. S. K. Das, A machine representation of finite T_o-topologies, Jour. of Assoc. Comp. Mach. 24 (1977), 676-692.

61. M. Davio, External solutions of unate Boolean equations, Philips Research Report 25 (1970), 201-206.

62. M. Davio and G. Bioul, Interconnection structure of injective counters composed entirely of JK flip-flops, Inf. and Cont. 33 (1977), 304-332.

63. M. Davio and J.-P. Deschamps, Classes of solutions of Boolean equations, Philips Research Report 24 (1969), 313-378.

64. M. Davio, J.-P. Deschamps, and J.-C. Lienard, Optimization of strictly non-blocking and rearrangeable interconnections, Philips Research Report 32 (1971), 266-296.

65. M. Davio and J.-J. Quisquater, Affine cascades, Philips Research Report 27 (1972), 109-125.

66. R. L. Davis, The number of structures of finite relations, Proc. of Amer. Math. Soc. 4 (1953), 486-495.

67. P. Delsarte and J.-J. Quisquater, Permutation cascades with normalized cells, Inf. and Cont. 23 (1973), 345-356.

68. R. S. Dembo and P. L. Hammer, A reduction algorithm for knapsack problems, Preprint, Univ. of Waterloo, Waterloo, Canada, 1977.

69. J. Denes, Az elektronikus szamologepek hehany nem szokvanyos felhasznalasa (Hungarian), KSH Ugyvitelgepesitesi Föosztaly Közlemenyei 1962, 17-54.

70. J. Denes, Az osszes n-edfoku permutacio eloallitasarol (Hungarian), Mat. Lapok 15 (1964), 239-241.

71. J. Denes, Bibliography of non-numerical applications of digital computers, Comp. Rev. 1968, 481-508.

72. J. Denes, Connections between transformation semigroups and graphs, Theorie des Graphes Journees International d'Etude, Rome, 1966, Dunod, Paris and Gordon and Breach, New York, 1967, 93-101. MR 36#2528.

73. J. Denes, On a problem of L. Fuchs, Acta Math. Szeged 23 (1962), 231-241. MR 27#1493.

74. J. Denes, On commutator subgroups of finite groups, Comm. Math. Univ. Santi Pauli 15 (1967), 61-65. MR 35#4293.

75. J. Denes, On graph representation of semigroups, Proc. of Calgary International Conf. on the Comb. Structures and Appl., Gordon and Breach, New York, 1970, 55-57.

76. J. Denes, On multiplication tables of finite quasigroups and semigroups, Preprint, Univ. of Surrey, London, 1967. MR 37#6391.

77. J. Denes, On some properties of commutator semigroups, Pub. Math. Debrecen 15 (1968), 283-285.

78. J. Denes, On some properties of commutator subgroups, Annales Univ. Sci. Budapest R. Eötvös, Sec. Math. 7 (1964), 123-127.

79. J. Denes, On transformations, transformation semigroups and graphs, Proc. of Colloq. on Graph Theory, Tihany, Hungary, 1966, Academic Press, New York and Akademiai Kiado, Budapest, 1968, 65-75. MR 38#3367.

80. J. Denes, The representations of a permutation as the product of a minimal number of transpositions and its connection with the theory of graphs, MTA Mat. Kut. Int. Közl. 4 (1959), 63-71. MR 22#6133.

81. J. Denes, Some combinatorial properties of transformations and their connection with the theory of graphs, Jour. of Comb. Theory 9 (1970), 108-116.

82. J. Denes, Transformations and transformation semigroups, Preprint, Univ. of Wisconsin, Madison, Wis., 1976.

83. J. Denes and A. D. Keedwell, Latin Squares and Their Applications, Akademiai Kiado, Budapest, Academic Press, New York, and English Univ. Press, London, 1974.

84. J. Denes, K. H. Kim, and F. W. Roush, A characterization of an inverse for semigroup elements, Seminar Report, Alabama State Univ., Montgomery, Al., 1976.

85. J. Denes, L. Pasztorniczky, and M. Szokolay, On some methods of data compression, Proc. of Fourth Colloq. on Microwaves Communications, Budapest, Hungary, 1970.

86. C. A. Desor and B. H. Whalen, A note on pseudoinverses, Jour. of SIAM Appl. Math. 11 (1963), 442-447.

87. H. M. Devadze, Generating sets and defining relations of certain semigroups of all binary relations in a finite set (Russian), XX Gercen Ctenija, Leningrad, 1967, 13-18.

88. H. M. Devadze, Generating sets of certain subsemigroups of the semigroup of all binary relations in a finite set (Russian), Ucenye Zap. Leningrad Gos. Ped. Inst. A. I. Gercen 387 (1968), 92-100.

89. H. M. Devadze, Generating sets of semigroups of all binary relations in a finite set (Russian), Doklady Akad. Nauk Belorus, SSSR 12 (1968), 765-768. MR 39#331.

90. H. M. Devadze, Relations defining the order in the ordered semigroup of all binary relations in a finite set (Russian), Izv. Vyss. Zaved. Mat. 3 (1968), 28-36. MR 37#2648.

91. R. P. Dilworth, A decomposition theorem for partially ordered sets, Ann. Math. 5 (1950), 161-166.

92. J. D. Dixon, The probability of generating the symmetric group, Math. Zeit. 110 (1969), 199-205.

93. M. P. Drazin, Pseudo-inverses in associative rings and semigroups, Amer. Math. Monthly 65 (1958), 506-514.

94. P. Dubreil, Contribution to the theory of semigroups (French), Mem. Acad. Sci. Inst. France 2(63), No. 3 (1941), 52 pp. MR 8#15.

95. A. L. Dulmage and N. S. Mendelsohn, Graphs and matrices, Graph Theory and Theoretical Physics, edited by F. Harary, Academic Press, London and New York, 1967, 167-277.

96. A. L. Dulmage and N. S. Mendelsohn, Structure of powers of nonnegative matrices, Canadian Jour. of Math. 17 (1965), 318-330.

97. P. Dwinger, Introduction to Boolean Algebra, Physica-Verlag, Würzburg, Germany, 1961.

98. D. O. Ellis and J. W. Gaddum, On solutions of systems of linear equations in a Boolean algebra, Bull. of Amer. Math. Soc. 56 (1950), 747.

99. P. Erdös, A. W. Goodman, and L. Posa, The representation of graphs by set intersections, Canadian Jour. of Math. 18 (1966), 106-111.

100. P. Erdös and D. J. Kleitman, Extremal problems among subsets of a set, Dis. Math. 8 (1974), 281-294.

101. P. Erdös and L. Moser, Problem Section, Combinatorial Theory and Applications III, Colloq. Math. Soc. Janos Bolyai, edited by A. Renyi and V. T. Sos, 4 (1970), 1159.

102. P. Erdös and A. Renyi, On random graphs I, Pub. Math. Debrecen 6 (1959), 290-297.

103. P. Erdös and A. Renyi, On random matrices II, Studia Sci. Math. Hungarica 3 (1968), 459-464.

104. P. Erdös and A. Renyi, On the evolution of random graphs, MTA Mat. Kut. Inst. Közl. 5 (1960), 17-61.

105. P. Erdös and J. Spencer, Probabilistic Methods in Combinatorics, Academic Press, New York and Akademiai Kiado, Budapest, 1974.

106. P. Erdös and P. Turan, Certain problems of statistical group theory, MTA III, Osztaly Kozl. 17 (1967), 51-57. MR 35#6744.

107. P. Erdös and P. Turan, On some problems of statistical group theory I, Zeitschr. für Warscheinlichtkeits Theorie und Verw. Gebiete 4 (1965), 175-186.

108. P. Erdös and A. Renyi, On some problems of a statistical group theory II, Acta Math. Hungarica 18 (1967), 151-163. MR 34#7624.

109. P. Erdös and A. Renyi, On some problems of a statistical group theory III, Acta Math. Hungarica 18 (1967), 309-320. MR 35#6743.

110. P. Erdös and A. Renyi, On some problems of a statistical group theory IV, Acta Math. Hungarica 19 (1968), 413-435. MR 38#1156.

111. M. Erne, Structure and enumeration formulas for topologies on a finite set (German), Manuscripta Math. 11 (1974), 221-259.

112. J. W. Evans, F. Harary, and M. S. Lynn, On the computer enumeration of finite topologies, Comm. of Comp. Mach. 10 (1967), 295-298.

113. P. C. Fishburn, The Theory of Social Choice, Princeton Univ. Press, Princeton, N. J., 1973.

114. D. G. Fitz-Gerald and G. B. Preston, Divisibility of binary relations, Bull. of Australian Math. Soc. 5 (1971), 75-86.

115. R. Freese, An application of Dilworth's lattice of maximal antichains, Dis. Math. 17 (1974), 107-109.

116. K. O. Friedrich, Functional Analysis, Lecture Notes, New York Univ., New York, 1950.

117. O. Frink, Jr., On the existence of linear algebras in Boolean algebras, Bull. of Amer. Math. Soc. 34 (1928), 329-333.

118. G. Frobenius, On Finite Groups (German), Sitzungs ber. Preuss Akad. Wiss. Berlin, 1885, 163-194.

119. L. Fuchs, Abelian Groups, Akademiai Kiado, Budapest, 1966.

120. L. Fuchs, Partially Ordered Algebraic Systems, Addison-Wesley, New York, 1963.

121. G. Gallo, P. L. Hammer, and B. Simeone, Quadratic knapsack problems, Preprint, Univ. of Waterloo, Waterloo, Canada, 1977.

122. M. B. Garman and M. I. Kamien, The paradox of voting: Probability calculations, Behavioral Sci. 13 (1968), 306-317.

123. Y. Give'on, Lattice matrices, Inf. and Cont. 7 (1964), 477-484.

124. L. M. Gluskin, Automorphisms of semigroups of binary relations (Russian), Uch. Zap. Gos. Univ. 6 (1967), 44-54.

125. L. M. Gluskin, Ideals of semigroups of transformations (Russian), Mat. Sbornik 47 (1959), 111-130.

126. R. L. Goodstein, Boolean Algebra, Macmillan, New York, 1963.

127. J. A. Green, On the structure of semigroups, Ann. Math. 54 (1951), 163-172.

128. C. Greene, Personal communication, 1977.

129. C. Greene and D. J. Kleitman, Proof techniques in the theory of finite sets, Studies in Combinatorics, edited by G.-C. Rota, Math. Assoc. of Amer., to appear.

130. C. Greene and D. J. Kleitman, Strong version of Sperner's theorem, Preprint, Mass. Inst. of Tech., Cambridge, Mass., 1976.

131. D. A. Gregory and N. J. Pullman, Prime Boolean matrices, a graph-theoretical approach, Preprint, Queen's Univ., Kingston, Canada, 1980.

132. G.-T. Guilbaud, Les theories de l'interet general et la problem logigue de l'agregation, Economie Appliequee 5 (1952), 501-584.

133. M. Hall, The Theory of Groups, Macmillan, New York, 1959.

134. P. Hall, On the representatives of subsets, Jour. of London Math. Soc. 10 (1935), 26-30.

135. P. R. Halmos, Lectures on Boolean Algebras, Van Nostrand, New York, 1963.

136. P. L. Hammer and S. Rudeanu, Boolean Methods in Operations Research, Springer-Verlag, Berlin and New York, 1968. MR 38#4132.

137. F. Harary, Graph Theory, Addison-Wesley, Reading, Mass., 1969.

138. F. Harary, The number of functional digraphs, Math. Ann. 138 (1959), 203-210. MR 22#18.

139. F. Harary, R. M. Karp, and W. T. Tutte, A criterion for planarity of the square of a graph, Jour. of Comb. Theory 2 (1967), 395-405.

140. F. Harary, R. Z. Norman, and D. Cartwright, Structural Models, John Wiley, New York, 1965.

141. F. Harary, E. M. Palmer, and R. C. Read, The number of ways to label structure, Psychometrika 32 (1967), 155-156.

142. F. Harary and R. C. Read, The probability of a given 1-choice structure, Psychometrika 31 (1966), 271-278.

143. D. W. Hardy and M. C. Thornton, The domain-range trace of a finite partial order, Preprint, Univ. of Nebraska, Lincoln, Neb., 1980.

144. D. W. Hardy and M. C. Thornton, Partial orders and their semigroups of closed relations, Preprint, Univ. of Nebraska, Lincoln, Neb., 1980.

145. D. W. Hardy and M. C. Thornton, Partial orders, closed relations and their automorphisms, Preprint, Univ. of Nebraska, Lincoln, Neb., 1980.

146. B. Harris, The asymptotic distribution of the order of elements in symmetric semigroups, Jour. of Comb. Theory 15 (1973), 66-74.

147. B. Harris and L. Schoenfeld, The number of idempotent elements in symmetric semigroups, Jour. of Comb. Theory 3 (1967), 122-135.

148. D. J. Hartfiel, On constructing nearly decomposable matrices, Proc. of Amer. Math. Soc. 27 (1971), 222-228.

149. D. J. Hartfiel, A simplified form for nearly reducible and nearly decomposable matrices, Proc. of Amer. Math. Soc. 24 (1970), 388-393.

150. D. J. Hartfiel and C. J. Maxson, A characterization of the maximal groups in B_X, Preprint, Texas A & M Univ., College Station, Texas, 1974.

151. M. B. Hedrick, Nearly Reducible and Nearly Decomposable Special Classes of Irreducible and Fully Indecomposable Matrices, Thesis, Univ. of Houston, Texas, 1972.

152. Z. Hedrlin, On the number of commuting transformations, Comm. Math. Univ. Carolinae 1963, 132-136. MR 36#2728.

153. L. Henkin, J. D. Monk, and A. Tarski, Cylindrical Algebra, Part I, North-Holland, Amsterdam and London, 1971.

154. A. J. Hoffman, An equation in matrices, Jour. of Comb. Theory 2 (1967), 393.

155. F. E. Hohn, Applied Boolean Algebra, Macmillan, New York, 1969.

156. F. E. Hohn and L. R. Schissler, Boolean matrices and the design of combinatorial relay circuits, Bell System Tech. Jour. 34 (1955), 177-202.

157. J. E. Hopcroft and J. D. Ullman, Formal Languages and Their Relation to Automata, Addison-Wesley, Reading, Mass., 1969.

158. J. M. Howie, The subsemigroup generated by the idempotents of a full transformation semigroup, Proc. of London Math. Soc. 41 (1966), 707-716.

159. E. Howorka, Generators for algebras of relations, Notices of Amer. Math. Soc. 24 (1971), A-4.

160. Y. Ijiri, On the generalized inverse of an incidence matrix, Jour. of SIAM Appl. Math. 13 (1965), 827-836.

161. K. Inada, A note on the simple majority decision rule, Econometrica 32 (1964), 525-531.

162. A. Jaeger, A plea for preordinators, Lecture Notes in Economics and Mathematical Systems, No. 141, Mathematical Economics and Game Theory, edited by R. Henn and O. Moeschlin, Berlin and New York, 1977, 605-615.

163. N. Jardine and R. Sibson, Mathematical Taxonomy, John Wiley, New York, 1971.

164. C. S. Johnson and F. R. McMorris, Completely cyclic injective semilattices, Proc. of Amer. Math. Soc. 36 (1972), 385-388.

165. K. Johnson, Real Representations of Finite Directed Graphs, Thesis, Univ. of Alabama, University, Ala., 1971.

166. C. Jordan, Treatise on Substitutions and Algebraic Equations (French), Gauthier-Villars, Paris, 1870.

167. H. Jürgensen and J. Loewer, Drawing Hasse diagrams of partially ordered sets, Preprint, Tech. Hoch. Darmstadt, Darmstadt, Germany, 1980.

168. H. Jürgensen and P. Wick, The semigroups of order ≤ 7 (German), Semigroup Forum 14 (1977), 69-79.

169. K. M. Kapp and H. Schneider, Completely 0-Simple Semigroups: An Abstract Treatment of the Lattice of Congruences, Benjamin, New York, 1969.

170. G. O. H. Katona, Extremal problems for hypergraphs, Math. Centre Tracts 56, edited by M. Hall and J. H. Van Lint, Amsterdam, 1974.

171. L. Katz, Probability of indecomposability of a random mapping function, Ann. Math. Stat. 26 (1953), 512-557. MR 17#48.

172. G. Keri, On the transformation of matrices to block triangular form by using row and column permutations, Preprint, Mathematical Institute, Hungarian Academy of Sciences, Budapest, 1977.

173. K. H. Kim, An extension of the Dulmage-Mendelsohn Theorem, Linear Alg. & Appl. 27 (1979), 187-197.

174. K. H. Kim, A Moore-Penrose inverse for Boolean relation matrices, Proc. of Second Australian Conf. on Comb. Math., Melbourne, Australia, 1973. Lecture Notes in Math., No. 403, Springer-Verlag, Berlin, Heidelberg and New York, 1974, 18-28. MR 50#6955.

175. K. H. Kim, New representation of posets, Proc. of International Colloq. on Infinite and Finite Sets, Keszthely, Hungary, 1973. Colloq. Math. Soc. Janos Bolyai 10 (1973), 241-250. MR 51#5422.

176. K. H. Kim, The number of open sets of finite topologies, Preprint, Pembroke State Univ., Pembroke, N. C., 1973.

177. K. H. Kim, The number of partially ordered sets, Jour. of Comb. Theory, Ser. B 13 (1972), 276-289.

178. K. H. Kim, On (0, 1)-Matrices, Lecture Notes, Instituto de Fisica e Matematica, Lisbon, Portugal, 1973. MR 54#2674.

179. K. H. Kim, The semigroup of Hall relations, Semigroup Forum 9 (1974), 253-260.

180. K. H. Kim, Subgroups of binary relations, Proc. of Second International Conf. on the Theory of Groups, Canberra, Australia, 1973. Lecture Notes in Math., No. 372, Springer-Verlag, Berlin, Heidelberg and New York, 1974, 188-196. MR 51#774.

181. K. H. Kim and J. R. Krabill, Abelian subsemigroups, enumeration, and universal matrices, Duke Math. Jour. 40 (1973), 587-598. MR 47#6922.

182. K. H. Kim and J. R. Krabill, Circulant Boolean relation matrices, Czech. Math. Jour. 24 (1974), 247-251.

183. K. H. Kim and G. Markowsky, Enumeration of finite topologies, Fourth Southeastern Conf. on Comb., Graph Theory and Comp., Boca Raton, Fl., 1973, 169-184. MR 50#3171.

184. K. H. Kim and G. Markowsky, The number of partially ordered sets, II, Jour. of Korean Math. Soc. 11 (1974), 7-17. MR 50#1991.

185. K. H. Kim and W. Maryland, Enumeration of regular topologies, Notices of Amer. Math. Soc. 22 (1975), A-336.

186. K. H. Kim, D. Rogers, and F. W. Roush, Connective relations on posets, Seminar Report, Alabama State Univ., Montgomery, Al., 1979.

187. K. H. Kim and F. W. Roush, The algebraic structure of the semigroup of binary relations, Glasgow Math. Jour. 22 (1981), 57-68.

188. K. H. Kim and F. W. Roush, Automata on one symbol, Preprint, Alabama State Univ., Montgomery, Al., 1980.

189. K. H. Kim and F. W. Roush, Clifford inverse semigroups and Boolean matrices, Proc. of Conf. on Semigroups in Honor of A. H. Clifford, Tulane Univ., New Orleans, 1979, 164-187.

190. K. H. Kim and F. W. Roush, Counting additive spaces of sets, Acta Sci. Math. Szeged 40 (1978), 81-87. MR 58#440.

191. K. H. Kim and F. W. Roush, Enumeration of normal topologies, Notices of Amer. Math. Soc. 24 (1977), A-519.

192. K. H. Kim and F. W. Roush, Generalization of Turan's theorem, Acta Math. Hungarica, to appear.

193. K. H. Kim and F. W. Roush, Generalized fuzzy matrices, Fuzzy Sets and Systems 4 (1980), 293-315.

194. K. H. Kim and F. W. Roush, Group-theoretic complexity, Jour. of Linear Alg. & Appl. 25 (1979), 289-297.

195. K. H. Kim and F. W. Roush, Idempotent fuzzy matrices, Seminar Report, Alabama State Univ., Montgomery, Al., 1981.

196. K. H. Kim and F. W. Roush, On generating regular elements in the semigroup of binary relations, Semigroup Forum 14 (1977), 29-32. MR 57#3289.

197. K. H. Kim and F. W. Roush, On inverses of Boolean matrices, Jour. of Linear Alg. & Appl. 22 (1978), 247-262.

198. K. H. Kim and F. W. Roush, On linear representations of semigroups of Boolean matrices, Proc. of Amer. Math. Soc. 63 (1977), 297-300.

199. K. H. Kim and F. W. Roush, On maximal submonoids of B_X, Semigroup Forum 15 (1977), 137-147. MR 57#3289.

200. K. H. Kim and F. W. Roush, On the Hamming distance between Boolean matrices, Jour. of Comb., Inf. & Sys. Sci. 3 (1978), 24-28.

201. K. H. Kim and F. W. Roush, On the Hamming distances, Seminar Report, Alabama State Univ., Montgomery, Al., 1977.

202. K. H. Kim and F. W. Roush, Orders of subsemigroups of T_n, Semigroup Forum 16 (1978), 203. MR 58#6005.

203. K. H. Kim and F. W. Roush, Posets and finite topologies, Pure & Appl. Math. Sci., to appear.

204. K. H. Kim and F. W. Roush, Quasi-varieties of binary relations, Acta Sci. Math. Szeged 40 (1978), 89-91.

205. K. H. Kim and F. W. Roush, Realizing all linear transformations, Jour. of Linear Alg. & Appl. 37 (1981), 97-101.

206. K. H. Kim and F. W. Roush, Seminar Notes on Boolean Matrices, Alabama State Univ., Montgomery, Al., 1979.

207. K. H. Kim and F. W. Roush, Some results on decidability of shift equivalence, Jour. of Comb., Inf. & Sys. Sci. 4 (1979), 123-146.

208. K. H. Kim and F. W. Roush, Two generator semigroups of binary relations, Jour. of Math. Psychology 17 (1978), 236-246.

209. K. H. Kim, F. W. Roush, and W. Schöenfeld, On the residual finiteness of free semigroups and composition of Boolean matrices, Proc. of Amer. Math. Soc. 63 (1977), 188.

210. K. H. Kim and S. Schwarz, The semigroup of circulant Boolean matrices, Czec. Math. Jour. 26 (1976), 632-635.

211. D. Klarner, The number of graded partially ordered sets, Jour. of Comb. Theory 6 (1969), 12-19.

212. D. J. Kleitman, Extremal properties of sets containing no two sets and their union, Preprint, Mass. Inst. Tech., Cambridge, Mass., 1977.

213. D. J. Kleitman, M. Edelberg, and D. Lubell, Maximal sized antichains in partial orders, Dis. Math. 1 (1971), 47-53.

214. D. J. Kleitman and B. L. Rothschild, Asymptotic enumeration of partial orders on a finite set, Tran. of Amer. Math. Soc. 205 (1975), 205-220.

215. D. J. Kleitman and B. L. Rothschild, The number of finite topologies, Proc. of Amer. Math. Soc. 25 (1970), 276-282.

216. D. J. Kleitman, B. L. Rothschild, and J. Spencer, The number of semigroups of order n, Proc. of Amer. Math. Soc. 55 (1976), 227-232.

217. D. König, Theory of Finite and Infinite Graphs (German), Chelsa, New York, 1950.

218. H. J. Kowalski, Lattice theoretic characterization of topological spaces (German), Math. Nach. 21 (1960), 297-318.

219. K. P. Kozlov, Concerning the existence of rigid cells in cell semigroups (Russian), Ucenye Zap. Leningrad Gos. Ped. Inst. A. I. Gercen 276 (1967), 131-133.

220. K. P. Kozlov, Concerning unrestricted varying cells in cell semigroups (Russian), Ucenye Zap. Leningrad Gos. Ped. Inst. A. I. Gercen 274 (1965), 143-149.

221. H. M. Krause, Group Structure and the Characteristics of Groups (German), E. Tech. Hochschule Zurich, Promotions Arbeit, Zurich, 1953. MR 15#99.

222. V. Krishnamurthy, On the number of topologies on a finite set, Amer. Math. Monthly 73 (1966), 154-157.

223. K. Krohn, J. Rhodes, and B. Tilson, Lectures on Finite Semigroups, Univ. of Calif., Berkeley, Calif., 1967.

224. H. D. Landahl and R. Runge, Outline of a matrix algebra for neural nets, Bull. of Math. Biophysics 8 (1946), 75-81.

225. R. S. Ledley, Boolean matrix equations in digital circuit design, IRE Tran. Elec. Comp. EC-8-2 (1959), 131-139.

226. R. S. Ledley, Digital methods in symbolic logic, Proc. of U. S. Nat. Acad. of Sci. 41-7 (1955), 498-511.

227. R. S. Ledley, Logic and Boolean algebra in medical science, Proc. of Conf. on Appl. of Undergrad. Math., Atlanta, Ga., 1973.

228. C. I. Lewis, A Survey of Symbolic Logic, Univ. of Calif. Press, Berkeley, Calif., 1918. Reprinted by Dover, New York, 1960.

229. E. S. Ljapin, Semigroups (Russian), Fitzmatgiz, Moscow, 1960. English translation by Amer. Math. Soc., Providence, R. I., 1968.

230. L. Lovasz, On covering of graphs, Proc. of Colloq. on Graph Theory, Tihany, Hungary, 1966. Academic Press, New York and Akademiai Kiado, Budapest, 1968, 231-236.

231. L. Löwenheim, On determinants (German), Math. Ann. 79 (1919), 222-236.

232. D. Lubell, A short proof of Sperner's theorem, Jour. of Comb. Theory 1 (1966), 299.

233. R. D. Luce, Connectivity and generalized cliques in sociometric group structure, Psychometrika 15-2 (1950), 169-170.

234. R. D. Luce, A note on Boolean matrix theory, Proc. of Amer. Math. Soc. 3 (1952), 382-388.

235. R. D. Luce, Semiorders and a theory of utility discrimination, Econometrica 24 (1956), 178-191.

236. R. D. Luce and A. D. Perry, A method of matrix analysis of group structure, Psychometrika 14-1 (1949), 95-116.

237. R. D. Luce and H. Raiffa, Games and Decisions, John Wiley, New York, 1957.

238. R. Mandel, Precise bounds associated with the subset construction on various classes of nondeterministic finite automata, Proc. of Seventh Annual Princeton Conf. on Inf. Sci., 1973, 263-267.

239. I. Malcev, Symmetric groupoids (Russian), Mat. Sbornik 31 (1952), 136-151.

240. W. W. Manning, Primitive Groups, Stanford Univ. Press, Stanford, Calif., 1921.

241. M. Marcus and H. Minc, A Survey of Matrix Theory and Matrix Inequalities, Prindle, Weber and Schmidt, Boston, Mass., 1964.

242. G. Markowsky, Bounds on the index and period of binary relation on a finite set, Semigroup Forum 13 (1976), 253-260.

243. G. Markowsky, Combinatorial Aspects of Lattice Theory with Applications to the enumerative of Free Distributive Lattices, Thesis, Harvard Univ., Cambridge, Mass., 1973.

244. G. Markowsky, Green's equivalence relations and the semigroup of binary relations on a set, Preliminary Report, Harvard Univ., Cambridge, Mass., 1972.

245. G. Markowsky, Idempotents and product representations with applications to the semigroup of binary relations, Semigroup Forum 5 (1972), 95-119.

246. W. Maryland, Connected Properties of Partially Ordered Sets, Thesis, Univ. of Alabama, University, Al., 1978.

247. E. J. McCluskey, Minimization of Boolean functions, Bell Sys. Tech. Jour. 35 (1956), 1417-1444.

248. M. C. McCord, Singular homology groups and homotopy groups of finite topological spaces, Duke Math. Jour. 33 (1966), 465-474.

249. R. McKinzie, The Representation of Relation Algebra, Thesis, Univ. of Colorado, Boulder, Col., 1966.

250. J. C. C. McKinsey, Postulates for the calculus of binary relations, Jour. of Sym. Log. 5 (1940), 85-97.

251. K. Menger, The algebra of functions, past, present, future, Rend. Mat. 20 (1961), 409-430.

252. L. D. Meshalkin, A generalization of Sperner's theorem and subsets of a finite sets, Preprint, 1977.

253. H. Minc, Nearly decomposable matrices, Jour. of Linear Alg. & Appl. 5 (1972), 181-187. MR 47#1832.

254. H. Minc, Nonnegative Matrices, Lecture Note, Technion-Israel Inst. of Tech., Haifa, Israel, 1974.

255. B. G. Mirkin, Analysis of Qualitative Attributes and Structure (Russian), Statistika, Moscow, 1980.

256. B. G. Mirkin, Group Choice, translated by Yelena Oliker, edited by Peter C. Fishburn, Winston and Sons, Washington, D. C., 1979.

257. J. S. Montague and R. J. Plemmons, Maximal subgroups of the semigroup of relations, Jour. of Alg. 13 (1969), 575-587.

258. E. H. Moore, General analysis, Part I, Mem. of Amer. Phil. Soc. 1 (1935).

259. M. Morse, Representation of geodesics, Amer. Jour. Math. 43 (1921), 33-51.

260. A. Mukhopadhyay, The square root of a graph, Jour. of Comb. Theory 2 (1967), 290-295.

261. M. Z. Nashed, Generalized Inverses and Applications, Academic Press, New York, 1975.

262. J. Neggers, On the number of isomorphism classes of certain types of actions and binary systems, Kyungpook Math. Jour. 16 (1976), 161-176.

263. J. Neggers, On the number of isomorphism classes of certain types of finite algebra, Kyungpook Math. Jour. 16 (1976), 141-147.

264. J. Neggers, Partially ordered sets and groupoids, Kyungpook Math. Jour. 16 (1976), 7-20.

265. J. Neggers, Representation of finite partially ordered sets, Jour. of Comb., Inf. & Sys. Sci. 3 (1979), 113-133.

266. J. Von Neumann, On regular rings, Proc. of U. S. Nat. Acad. Sci. 22 (1936), 296-300.

267. G. N. De Oliveira, Binary square roots of matrices, Tech. Report, No. 14, Univ. Fed. de Pernambuco, Recife, Brazil, 1971.

268. J.-P. Olivier and D. Serrato, Dedekind categories and correspondences in classical and other categories (French), Preprint, Univ. Paul Valery, Montpellier, France, 1980.

269. O. Ore, Graph and Correspondence, Festschift Zum 60, Geburtstag Von Prof. Andreas Speiser, Zurich, Orell Fussi Verlag, 1945, 184-191. MR 7#215.

270. O. Ore, Some remarks on commutators, Proc. of Amer. Math. Soc. 2 (1951), 307-314.

271. O. Ore, Theory of Graphs, Amer. Math. Soc., Providence, R. I., 1962.

272. G. Pappy, Groupoids (French), Presses Universitaires, Paris, 1965.

273. W. Parry, Entropy and Generators in Ergodic Theory, Benjamin, New York, 1964.

274. W. Parry, Intrinsic Markov chains, Tran. of Amer. Soc. 112 (1964), 55-66.

275. K. R. Parthasarathy, Enumeration of paths in digraphs, Psychometrika 29 (1964), 153-165.

276. C. S. Peirce, On the algebra of logic, Amer. Jour. Math. 3 (1880), 15-57.

277. R. Penrose, A generalized inverse for matrices, Proc. of Camb. Phil. Soc. 5 (1955), 406-413.

278. M. Petrich, Introduction to Semigroups, Merrill, Columbus, Ohio, 1973.

279. M. Petrich, Representations of semigroups and the translational hull of a regular Rees matrix semigroup, Tran. of Amer. Math. Soc. 143 (1969), 303-318.

280. M. Petrich, Topics in Semigroups, Lecture Note, Penn. State Univ., University Park, Pa., 1967.

281. M. Petrich, Translational hull and semigroups of binary relations, Glasgow Math. Jour. 9 (1968), 1-11.

282. R. J. Plemmons, Generalized inverses of Boolean relation matrices, Jour. of SIAM Appl. Math. 20 (1971), 426-433.

283. R. J. Plemmons, Personal communication, 1975.

284. R. J. Plemmons, Regular nonnegative matrices, Tech. Report, Centre des Researches Math. Univ. de Montreal, 1972.

285. R. J. Plemmons and B. M. Schein, Groups of binary relations, Semigroup Forum 1 (1970), 267-271.

286. R. J. Plemmons and M. T. West, On the semigroup of binary relations, Pacific Jour. Math. 35 (1970), 143-153.

287. L. M. Popova, Defining relations of certain semigroups of partial transformations of a finite set (Russian), Ucenye Zap. Leningrad Gos. Ped. Inst. A. I. Gercen 218 (1961), 191-212.

288. L. V. Potemkin, Semigroup rings of symmetric semigroups of binary relations (Russian), XIIth All Soviet Union Algebraic Colloq., Sverdovsk, 1973, 226.

289. S. B. Presic, Une classe d'equations matricielles et l'equation fonctionelle $f^2 = f$ (French), Pub. Inst. Math. (Beograd), 8 (1968), 143-148.

290. G. B. Preston, Inverse semi-groups, Jour. of London Math. Soc. 29 (1954), 396-403.

291. N. J. Pullman, A property of infinite products of Boolean matrices, Jour. of SIAM Appl. Math. 15-4 (1967), 871-873.

292. W. V. Quine, The problem of simplifying truth functions, Amer. Math. Monthly 59 (1952), 521-531.

293. I. Rabinovitch, The Scott-Suppes theorem on semiorders, Jour. of Math. Psychology 15 (1977), 209-212.

294. C. R. Rao, Generalized inverse of Boolean valued matrices, Lecture delivered at the Conf. on Comb. Math., Univ. of Delhi, Delhi, India, 1972.

295. C. R. Rao and S. K. Mitra, Generalized Inverse of Matrices and Its Applications, John Wiley, New York, 1971.

296. P. P. Rao, Generalized Inverse of Special Types of Matrices, Thesis, Indian Stat. Inst., Calcutta, India, 1974.

297. P. P. Rao and K. B. Rao, On generalized inverse of Boolean matrices, Tech. Report, Indian Stat. Inst., Calcutta, India, 1973.

298. R. C. Read, A note on the number of functional digraphs, Math. Ann. 143 (1961), 109-110. MR 22#10919.

299. L. Redei, Theory of Finitely Generated Commutative Semigroups (German), Physica Verlag, Wüzburg, 1963. (English translation edited by N. Reilly, The Theory of Finitely Generated Commutative Semigroups, Pergamon Press, Oxford and New York, 1965).

300. A. Renyi, Connected graphs, Part I, MTA Mat. Kut. Int. Közl. 4 (1959), 385-388. MR 23#4136.

301. A. Renyi, Some remarks on the theory of trees, MTA Mat. Kut. Int. Közl. 4 (1959), 73-85. MR 22#6735.

302. A. H. Rhemtulla, On a problem of L. Fuchs, Studia Sci. Math. Hungarica 4 (1969), 195-200. MR 30#1468.

303. J. Rhodes, Finite binary relations have no more complexity than finite functions, Semigroup Forum 7 (1974), 92-103.

304. J. Rhodes, The fundamental lemma of complexity for arbitrary finite semigroups, Bull. of Amer. Math. Soc. 7 (1968), 1104-1109.

305. J. Rhodes, A proof of the fundamental lemma of complexity (strong version) for arbitrary finite semigroups, Preprint, Univ. of Calif., Berkeley, Calif., 1975.

306. J. Rhodes, A proof of the fundamental lemma of complexity (weak version) for arbitrary finite semigroups, Jour. of Comb. Theory, Ser. A 10 (1971), 22-73.

307. J. Rhodes, Some results on finite semigroups, Jour. of Alg. 4 (1966), 471-504.

308. J. Rhodes, The structure of finite semigroups, Preprint, Univ. of Calif., Berkeley, Calif., 1975.

309. J. Rhodes and B. Tilson, Lower bounds for complexity of finite semigroups, Jour. of Pure and Appl. Alg. 1 (1971), 79-95.

310. D. J. Richman and H. Schneider, Primes in the semigroup of non-negative matrices, Linear & Multiplicative Alg. 2 (1974), 135-140.

311. J. Riguet, Binary relations, closure, and Galois correspondences (French), Bull. Soc. Math. France 76 (1948), 114-115.

312. J. Riguet, On the set of regular binary relations (French), C. R. Acad. Sci. Paris 231 (1950), 936.

313. J. Riguet, Some applications of the theory of relations (French), Collect. Log. Math. Paris A5 (1954), 141-144.

314. W. H. Riker and P. C. Ordeshook, An Introduction to Positive Political Theory, Prentice-Hall, Englewood Cliffs, N. J., 1973.

315. J. Riordan, Introduction to Combinatorial Analysis, John Wiley, New York, 1958.

316. F. Robert, Discrete Iterations, A Metric Study (French), Lecture Note, Universite Scientifique et Medicale de Grenoble, Grenoble, France, 1981.

317. D. G. Rogers, Connective relations on linearly ordered sets, Preprint, Oxford Univ., Oxford, England, 1977.

318. D. G. Rogers, Similarity relations on finite ordered sets, Jour. of Comb. Theory, Ser. A 23 (1977), 88-98.

319. A. N. Rosenberg, On k-potent Boolean matrices, Seminar Report, Alabama State Univ., Montgomery, Al., 1977.

320. D. Rosenblatt, Aggregation in matrix models of resources flows I: Boolean relation matrix methods, Amer. Stat. June, 1967, 32-37.

321. D. Rosenblatt, On the graphs and asymptotic forms of finite Boolean relation matrices, Naval Res. Log. Quart. 4 (1957), 151.

322. D. Rosenblatt, On the graphs of finite idempotent Boolean relation matrices, Jour. Res. Nat. Bureau of Standards, B-Math. & Physics 61B-4 (1963), 249-256.

323. P. H. Rossi, Power and community structure, Midwest Jour. of Pol. Sci. 4 (1960), 390-401.

324. J. Rosta, On an extension of Cayley's theorem (German), Math. Nach. 41 (1969), 223-226.

325. J. Rothenberg, The Measurement of Social Welfare, Prentice-Hall, Englewood Cliffs, N. J., 1961.

326. N. Rouche, Some properties of Boolean equations, IRE Tran. Elec. Comp., EC-7 (1958), 291-298.

327. S. Rudeanu, Boolean Functions and Equations, North-Holland, Amsterdam, London and New York, 1974.

328. S. Rudeanu, On Boolean matrix equations, Rev. Roumaine Math. Pures. Appl. 17 (1972), 1075-1090.

329. D. E. Rutherford, The eigenvalue problem for Boolean matrices, Proc. of Royal Soc. Edinburgh A-67 (1965), 25-28.

330. D. E. Rutherford, Inverse of Boolean matrices, Proc. of Glasgow Math. Assoc. 6 (1963), 49-53.

331. H. J. Ryser, Combinatorial Mathematics, John Wiley, New York, 1963.

332. H. J. Ryser, A generalization of the matrix equation $A^2 = J$, Jour. of Linear Alg. & Appl. 3 (1970), 451-460.

333. B. M. Schein, A construction for idempotent binary relations, Proc. of Japan Acad. 46 (1970), 246-247.

334. B. M. Schein, Homomorphisms of symmetric semigroups into inverse semigroups (Russian), Proc. of Theory of Semigroups and Its Appl., Saratov Univ., Saratov, U.S.S.R., 2 (1971), 90-93.

335. B. M. Schein, Personal communications, 1973.

336. B. M. Schein, Regular elements of the semigroup of all binary relations, Semigroup Forum 13 (1976), 95-102.

337. B. M. Schein, Relation algebras and function semigroups, Semigroup Forum 1 (1970), 1-62.

338. B. M. Schein, Semigroups of binary relations (Russian), Doklady Akad. Nauk. SSSR 65 (1965), 1011-1014.

339. B. M. Schein, Semigroups of binary relations, Proc. of Mini-Conf. on Alg. Semigroup Theory, Szeged, Hungary, 1972.

340. J. A. Schreider, Equality, Resemblance, and Order, Mir Publishers, Moscow, 1975.

341. E. Schröder, Vorlesungen Uber Die Algebra der Logik (German), Band III, Algebra und Logik der Relative I, Leipzig, Teubner, 1895.

342. S. Schwarz, Circulant Boolean matrices, Czech. Math. Jour. 24 (1974), 252-253.

343. S. Schwarz, A counting theorem in the semigroup of circulant Boolean matrices, Czech. Math. Jour. 27 (1977), 504-510.

344. S. Schwarz, On some semigroups in combinatorics, Proc. of Mini-Conf. on Alg. Semigroup Theory, Szeged, Hungary, 1972, 24-31.

345. S. Schwarz, On the semigroup of binary relations on a finite set, Czech. Math. Jour. 20 (1970), 632-679.

346. S. Schwarz, The semigroup of fully indecomposable relations and Hall relations, Czech. Math. Jour. 23 (1973), 151-163.

347. S. Schwarz, Zur Theorie der Halbgruppen (German), Sbornik Prac Prirodovedekej Fakulty Slovenskej Univ. v, Bratislava, No. 6, 1943.

348. B. Schweizer and A. Sklar, The algebra of functions, Math. Ann. 139 (1960), 366-382.

349. B. Schweizer and A. Sklar, The algebra of functions II, Math. Ann. 143 (1961), 440-447.

350. B. Schweizer and A. Sklar, The algebra of functions III, Math. Ann. 161 (1965), 171-196.

351. B. Schweizer and A. Sklar, Function systems, Math. Ann. 172 (1967), 1-16.

352. D. Scott, Measurement structures and linear inequalities, Jour. of Math. Psychology 1 (1964), 233-247.

353. D. Scott and P. Suppes, Foundational aspects of theories of measurement, Jour. of Sym. Log. 23 (1958), 113-128.

354. J. E. Scroggs and P. L. Odell, An alternate definition of a pseudoinverse of a matrix, Jour. of SIAM Appl. Math. 14 (1966), 795-810.

355. W. Semon, The Application of Matrix Methods in the Theory of Switching, Thesis, Comp. Lab., Harvard Univ., Cambridge, Mass., 1954.

356. A. K. Sen, Collective Choice and Social Welfare, Holden-Day, San Francisco, 1970.

357. A. K. Sen, A possibility theorem on majority decision, Econometrica 34 (1966), 491-499.

358. C. E. Shannon, A symbolic analysis of relay and switching circuits, Tran. of Amer. Inst. of Elec. Eng. 57 (1938), 713-723.

359. L. Shapley, Political science, Lecture Note: Math. in Behavioral Sci., Williams College, Williamstown, Mass., 1973.

360. H. Sharp, Cardinality of finite topologies, Jour. of Comb. Theory 5 (1968), 82-86.

361. H. Sharp, Quasi orderings and topologies on finite sets, Proc. of Amer. Math. Soc. 17 (1966), 1344-1349.

362. R. D. Sheffield, On pseudoinverses of linear transformations in Banach spaces, Oak Ridge Nat. Lab., Report 2133, Oak Ridge, Tenn., 1956.

363. L. N. Shervin, The Sverdlovsk tetrad, Semigroup Forum 4 (1972), 274-280.

364. A. Shimbell, Applications of matrix algebra to communication nets, Bull. of Math. Biophysics 13 (1951), 165-178.

365. A. Shimbell, Structural parameters of communication nets, Bull. of Math. Biophysics 15 (1953), 501-507.

366. Z. Shmuely, The lattice of idempotent binary relations, Preprint, 1977.

367. E. G. Shtov, Homomorphisms of all semigroup of all partial transformations (Russian), Izv. Vys. Uch. Zav. 33-22 (1961), 177-184.

368. R. Sikorski, Boolean Algebra, Third Edition, Springer-Verlag, Berlin, Heidelberg and New York, 1967.

369. R. J. Simpson, The Application of Rectangular Relations to the Study of Binary Relations on a Set, Thesis, Univ. of Tenn., Knoxville, Tenn., 1972.

370. R. Sinkhorn and P. Knopp, Problems involving diagonal products in nonnegative matrices, Tran. of Amer. Math. Soc. 136 (1969), 67-75.

371. S. Smale, Diffeomorphism with many periodic points, Differential and Combinatorial Topology, Princeton Univ. Press, Princeton, N. J., 1965, 63-80.

372. J. Spencer, Turan's theorem for k-graphs, Dis. Math. 2 (1972), 183-186.

373. R. P. Stanley, On the number of open sets of finite topologies, Jour. of Comb. Theory, Ser. A 10 (1971), 74-79.

374. R. P. Stanley, Ordered Structures and Partitions, Mem. of Amer. Math. Soc., Providence, R. I., 1972.

375. D. Stephen, Topology on finite sets, Amer. Math. Monthly 75 (1968), 739-741.

376. P. Stiffler, Extension of the fundamental theorem of finite semigroups, Preprint, 1974.

377. R. E. Stong, Finite topological spaces, Tran. of Amer. Math. Soc. 123 (1966), 325-340.

378. P. Suppes and J. Zinnes, Basic measurement theory, edited by R. D. Luce, R. R. Bush and E. Galanter, Handbook of Mathematical Psychology, John Wiley, New York, 1963, 1-76.

379. A. Suskevitch, Theory of Generalized Groups (Russian), Gos. Nauk-Tekl. Izd. Kharkov, U.S.S.R., 1920.

380. M. Szalay, Number of Theoretical Extremal Problems (Hungarian), Thesis, Eötvös Lorand Univ., Budapest, 1973.

381. M. Tchunte, On the Decomposition of Boolean Matrices, Lecture Note, Univ. of Grenoble, Grenoble, France, 1980.

382. G. Thierrin, On the inversive elements and the idempotent elements of an inverse semigroup (French), C. R. Acad. Sci. Paris, 24 (1954), 33-34.

383. W. J. Thron, Lattice-equivalence of topological spaces, Duke Math. Jour. 29 (1962), 671-680.

384. B. Tilson, Decomposition and complexity of finite semigroups, Semigroup Forum 3 (1971), 189-250.

385. K. Z. Todorov, Directed graphs and matrices, Preprint, Bulgarian Academy of Sciences, Sofia, 1976.

386. E. Török, On the representation of nth grade mapping (Hungarian), Mat. Lopok 19 (1968), 143-146. MR 39#2895.

387. P. Turan, On some phenomena in the theory of partitions, Jour. of Arit. Bordeaux 24-25 (1975), 311-319.

388. V. V. Vagner, Generalized groups (Russian), Doklady Akad. Nauk SSSR N. S. 84 (1952), 1119-1122.

389. A. Varga, P. Ecsedi-Toth, and F. Moricz, An effective method for minimizing Boolean polynomials, Tech. Report, Szeged State Univ., Szeged, Hungary, 1977.

390. N. N. Vorob'ev, Defect ideals of associative system of substitutions (Russian), Ucenye Zap. Leningrad Gos. Univ. Sec. Mat. 16 (1949), 126-134.

391. N. N. Vorob'ev, Normal subsystems of a finite associative system (Russian), Doklady Akad. Nauk SSSR 58 (1947), 1877-1879. MR 9#330.

392. N. N. Vorob'ev, On canonical representations of elements of symmetric associative systems (Russian), Ucenye Zap. Leningrad Gos. Univ. A. I. Gercen 103 (1955), 75-82.

393. N. N. Vorob'ev, On ideals of associative systems (Russian), Ucenye Zap. Leningrad Gos. Univ. A. I. Gercen 89 (1953), 161-166. MR 17#943.

394. N. N. Vorob'ev, On the theory of ideals of associative systems (Russian), Ucenye Zap. Leningrad Gos Univ. A. I. Gercen 103 (1955), 31-73.

395. A. D. Wallace, Relation Theory, Lecture Note, Univ. of Florida, Gainesville, Fla., 1963.

396. B. Ward, Majority voting and alternate forms of public enterprises, edited by J. Margolis, Public Economy of Urban Communities, Johns Hopkins Univ. Press, Baltimore, Md., 1965.

397. J. H. M. Wedderburn, Boolean linear associative algebra, Ann. Math. 35 (1934), 185-194.

398. H. C. White, S. A. Boorman, and R. L. Breiger, Social structure from multiple networks, I: Blockmodels, roles, and positions, Amer. Jour. of Sociology 81 (1976), 730-780.

399. A. N. Whitehead and B. Russell, Principia Mathematica, Vol. 1, Cambridge, England, 1925.

400. A. Wieczorek, On representations of social preferences an algebraic approach, Lecture Notes in Economics and Mathematical Systems, No. 141, Mathematical Economics and Game Theory, edited by R. Henn and O. Moeschlin, Heidelberg and New York, 1977, 234-249.

401. R. F. Williams, Classification of subsets of finite type, Ann. Math. 2-99 (1974), 380-381.

402. R. F. Williams, Classification of symbol spaces of finite type, Bull. of Amer. Math. Soc. 77 (1971), 7439-7443.

403. E. S. Wolk, A characterization of partial order, Bull. de L'academie Polonaise des Sci., Ser. des Sci. Math. Astr. et Phy. 17 (1969), 207-208.

404. J. Wright, Cycle Indicators of Certain Classes of Types of Quasi-Orders of Topologies, Thesis, Univ. of Rochester, Rochester, New York, 1972.

405. J. Yackel, Inequalities and asymptotic bounds for Ramsey numbers, Jour. of Comb. Theory, Ser. B 13 (1972), 56-68.

406. K. Yamamoto, Logarithmic order of free distributive lattices, Jour. of Japan Math. Soc. 6 (1954), 343-353.

407. J.-C. Yang, A theorem on the semigroup of binary relations, Proc. of Amer. Math. Soc. 22 (1969), 134-135.

408. L. A. Zadeh, Fuzzy sets, Inf. & Cont. 8 (1965), 338-353.

409. K. A. Zaretski, Abstract characteristic of the semigroup of all binary relations (Russian), Leningrad Gos. Ped. Inst. Ucenye Zap. 183 (1958), 252-263.

410. K. A. Zaretski, Abstract characteristic of the semigroup of all reflexive binary relations (Russian), Leningrad Gos. Ped. Inst. Ucenye Zap. 183 (1958), 265-269.

411. K. A. Zaretski, Maximal submonoids of the semigroups of binary relations, Semigroup Forum 9 (1974), 196-206.

412. K. A. Zaretski, Regular elements in the semigroup of binary relations (Russian), Uspeki Mat. Nauk 17-3 (1962), 105-108.

413. K. A. Zaretski, The semigroup of binary relations (Russian), Sbornik 61 (1963), 291-305.

INDEX